WAR AS BUSINESS

War as Business
Technological Change and Military Service Contracting

ARMIN KRISHNAN
Salford Business School, University of Salford, UK

ROUTLEDGE
Routledge
Taylor & Francis Group
LONDON AND NEW YORK

First published 2008 by Ashgate Publishing

Published 2016 by Routledge
2 Park Square, Milton Park, Abingdon, Oxon OX14 4RN
711 Third Avenue, New York, NY 10017, USA

Routledge is an imprint of the Taylor & Francis Group, an informa business

British Library Cataloguing in Publication Data
Krishnan, Armin
War as business : technological change and military service contracting
1. Defense industries 2. Defense contracts 3. Contracting out 4. Military supplies - Technological innovations
I. Title
338.4'7355

Library of Congress Cataloging-in-Publication Data
Krishnan, Armin, 1975-
War as business : technological change and military service contracting / by Armin Krishnan.
p. cm.
Includes bibliographical references and index.
ISBN 978-0-7546-7167-1
1. War--Economic aspects--United States. 2. Technology--Military aspects. 3. Defense contracts--Economic aspects--United States. 4. Defense industries--Technological innovations--Economic aspects--United States. 5. Security, International. I. Title.
HB195.K7 2007
338.4'7355--dc22

2007025288

ISBN 13: 978-0-7546-7167-1 (hbk)

Contents

List of Tables and Figure

Abbreviations

ABL	Airborne Laser
APC	Armoured Personnel Carrier
AWACS	Airborne Warning an Control System
BICC	Bonn International Center for Conversion
BMD	Ballistic Missile Defense
C4I	Command, Control, Communications, Computers and Intelligence
DARPA	Defense Advanced Research Projects Agency
DHS	Department of Homeland Security
DLO	Defence Logistics Organisation (UK)
DoD	Department of Defense (US)
EADS	European Aeronautic Defense and Space Company
FAS	Federation of American Scientists
FCS	Future Combat Systems (US)
FFRDC	Federally Funded Research and Development Centers
GAO	General Accounting Office (US)
GPS	Global Positioning System
ISR	Intelligence, Surveillance, Reconnaissance
IW	Information Warfare
JCS	Joint Chiefs of Staff
JSF	Joint Strike Fighter
JSTARS	Joint Surveillance Target Attack Radar System
KBR	Kellogg Brown and Root
LIC	Low Intensity Conflict
LM	Lockheed Martin
LOGCAP	Logistics Civilian Augmentation Program
MBT	Main Battle Tank
MIC	Military-Industrial Complex
MoD	Ministry of Defence (UK)
MOUT	Military Operations in Urban Terrain
M&S	Modelling and Simulation
NAO	National Audit Office (UK)
NCW	Network-Centric Warfare
OEF	Operation Enduring Freedom
OIF	Operation Iraqi Freedom
OOTW	Operations Other Than War
PEO STRI	Program Executive Office Simulation, Training, and Instrumentation (US)
PFI	Private Finance Initiative

PGM	Precision Guided Munitions
PMC	Private Military Company
PMF	Private Military Firm
PPP	Private Public Partnership
QDR	Quadrennial Defense Review
R&D	Research and Development
RBA	Revolution in Business Affairs
RMA	Revolution in Military Affairs
RML	Revolution in Military Logistics
RPG	Rocket Propelled Grenade
SBA	Simulation Based Acquisition
SDR	Strategic Defense Review (UK)
SIGINT	Signals Intelligence
SIPRI	Stockholm International Peace Research Institute
SS&M	Stockpile Stewardship & Management
TMD	Theater Missile Defense
UAV	Unmanned Aerial Vehicle
UCAV	Unmanned Combat Aerial Vehicle
WMD	Weapons of Mass Destruction

Acknowledgements

I want to thank the University of Salford, which funded me and enabled me to conduct my research on military privatization. I am very grateful to the European Studies Research Institute, especially to Professor Keiger, and for the level of support I received. I am very much indebted to Nikolas Gardner and Larry Valero, who supplied me with many ideas and who gave me important feedback on my work. I would also like to thank Keith Hayward and Alaric Searle for reading the initial version of the manuscript and for their constructive criticism. There are many others, in particular my colleagues and fellow researchers in the School of English, Sociology and Contemporary History, who helped me finding my way during the sometimes challenging time at Salford. My sincere thanks to all of them. Finally, I need to thank my friend Ian, who gave me many insights into the world of computer simulations.

Introduction

A quiet revolution is transforming the military world and its implications are hard to predict at the moment. Technology shapes the conduct of war in a tremendous way. It has always done this and there is little new about it. Everything in modern war seems to revolve around technology: it determines how wars are fought, how armies are organized and also many of the limitations they have.[1] This dominance of technology, which is so characteristic of modern times, does not imply that technological superiority can always win battles and decide wars. There is plenty of evidence to the contrary. Nevertheless, the belief in technology as a solution and path to victory remains unbroken.

This is indicated by the claim that a Revolution in Military Affairs (RMA) triggered by information technology could change warfare so radically that it could break with the continuity of (military) history. Whoever can master this revolution, so the claim goes, could have a decisive advantage over one's enemies.[2] In the end it might be irrelevant whether this will be the case or not, as there might not be enough clear test cases to make a verdict. Anyway proponents of this idea are likely to claim in face of ambiguous success that the true revolution was not yet achieved and that it lies shortly in the future. Therefore it can be expected that the current trend towards high-tech warfare will persist and that most modern armed forces will push military technology to its limits with ever increasing sophistication and complexity of weapons systems and equipment.

The result of this will be, as can already be seen in the development of the past decades, an ever increasing reliance on private companies, not only for developing and manufacturing these military systems, but increasingly also for maintaining,

1 Martin van Creveld argues in the introduction of his bock *Technology and War* that 'The present volume rests on one very simple premise which serves as its staring point, argument, and *raison d'être* rolled into one. It is that war is completely permeated by technology and governed by it. The causes that lead to wars, and the goals for which are they fought; the blows with which campaigns open, and the victories with which they (sometimes) end; the relationship between the armed forces and the societies that they serve; planning, preparation, execution, and evaluation; operations and intelligence and organization and supply; objectives and methods and capabilities and missions; command and leadership and strategy and tactics; even the very conceptual frameworks employed by our brains in order to think about war and its conduct – not one of these is immune to the impact that technology has had and does have and will always have.' (van Creveld 1989, 1).

2 Clifford Rogers has coined this phrase and he claims that 'To my mind, an RMA is simply a Revolutionary change in how war is fought – a change that can often be recognized by the ease with which "participating" armed forces can defeat "nonparticipating" ones.' (Rogers 2000, 22).

supporting and operating them (Markusen 2001, 4). The most modern armed forces in particular are no longer able to operate effectively without a private 'army of contractors', which keeps troops, weapons and equipment operational (Cahlink 2002, 43). In many cases military contractors do not just enable soldiers to achieve their missions, but they in a sense create and provide military capabilities, which would not exist without them. Increasing technological complexity therefore drives military privatization. This has already raised serious questions regarding what kind of responsibility should be given to private actors in respect to national and international security. Military privatization touches the core of the modern concept of sovereignty: the right and ability of a country to defend itself and to use military force as it deems necessary. By granting private companies such an extensive role in the military sphere, states might create severe limitations for using military force and might even create dangers for themselves. The quiet revolution mentioned above is not about the RMA itself, which has already produced a tremendous body of literature and is now at the focus of most works on contemporary and future warfare. The quiet revolution mentioned is the rarely noticed infiltration of private interests in the realm of warfare.

This book looks at technical services provided by private companies to national armed forces. The book falls into the area of security studies, as it looks at the creation and employment of military force from an essentially non-military angle. It is also a contribution to strategic and defence studies, as it covers the use of armed force and the problem of defence procurement and its implications. This study is not concerned with whether or not technology can win wars, or how future wars are fought and won. This can be left to the vast amount of speculative literature on future warfare. It is also not a criticism of the RMA literature, but meant as an addition to it, as it deals with a largely ignored and, at the same time, immensely important aspect of it.

This study draws upon a wide range of literature. From the broad field of international relations originates a growing literature on the privatization of security and warfare. The bulk of it deals with the phenomenon of the emergence of private authority and the new actors in warfare, such as militias, insurgents, warlords, transnational criminal organizations and mercenaries (Hall and Biersteker 2002). In many respects a classic is Martin van Creveld's *Transformation of War*, in which he claims that future warfare will resemble warfare in the Middle Ages and that modern armed forces (and with them nation states) are becoming irrelevant (van Creveld 1991; van Creveld 1999). There is now a whole subtype of literature on *New Wars* and *Fourth Generation Warfare* (Kaldor 1999; Münkler 2005; Hammes 2004). Very little of it is relevant for this study, however, which deals with what can be called *military privatization*. Military privatization includes the privatization of defence assets and the outsourcing of all kinds of services from the armed forces to the private sector. At the moment there still is not much literature available on the phenomenon of military privatization. A non-academic, but nevertheless very useful introduction is given by Ken Silverstein in his *Private Warriors*, where he describes the inner-workings of the US military-industrial complex of the 1990s (Silverstein 2000). Robert Mandel explains the complexity of *The Privatization of Security* in his book and the many conceptual obstacles for scholars of studying the phenomenon (Mandel 2002). In some sense Peter Singer's *Corporate Warriors* is groundbreaking in its attempts to outline a new global military industry (Singer 2003). Peter Singer mentions technical services in his book at various points (Singer 2003, 62-64, 100, 136-148, 156, 160, 172-174), but altogether they appear to be a rather marginal aspect within the overall military industry. More recently Deborah Avant has published a book on *The Market for Force*, in which she looks at the problem of controlling the use of private force and the implications for sovereignty (Avant 2005). The three studies (Mandel's, Singer's and Avant's) predominantly deal with private security providers and non-technical services. The main problem with the literature on the privatization of warfare is that it focuses too much on rather marginal aspects of it, such as private security companies operating in war zones and the 'new mercenarism'. The available literature falls short developing an overall theory why armed forces have begun to outsource technical military services.

What is also needed is an overall conceptual framework, which can include most of the relevant phenomena and a theory that explains how they are connected. The solution proposed in this book is to look broadly at services rather than particular company types or industries. Then it becomes apparent that the origins of the current military privatization can be found in the growing complexity of military equipment. The hypothesis put forward in this study is that technological change is at the root of the present drive towards privatization. Of course, other studies, most importantly Peter Singer's work, have given a whole catalogue of reasons for military privatization (technological change being only one of them), and each of the reasons given seems to be important, if looked at individually. So why claim that technology is decisive in the whole process? This assumption is based on three facts:

- The technologically most advanced armed forces, the American and the British forces, have the highest degree of privatization, while those armed forces, which employ less advanced technology, tend to be far less dependent on the private sector.[3]
- Even technologically advanced countries, where there is a much stronger trust in the public sector, like France and Germany, have begun to privatize their military in the face of their present RMA inspired modernization programmes.[4]

3 An important example is China, which is one of the few countries in the world that has a completely state-owned defence industry. As a result, China is far behind other great powers in terms of military technology and the only area where the gap is not quite as wide is ballistic missile technology and only because of its massive investments. The greatest impediments to China's pursuit of the RMA are its lack of access to foreign military technology and its sclerotic and overly centralized defence establishment. A successful modernization of the Chinese armed forces might require a change of the rigid political system (Shambaugh 2002, 243-250; Ji 2004, 97-117).

4 Germany has begun to privatize military training and army logistics. France has recently begun to privatize military flight training and the government has announced that it will give greater responsibility to the private sector. See below chapters 4 and 6.

- Even the outsourcing of services, which appear to be not technical in nature, like military training or logistics/management, turn out to be highly technical after a closer look.[5]

Previous works on the current privatization of warfare have almost exclusively focused on the rise of nonstate violence in form of militias, warlords, international criminal organizations, mercenaries and transnational security companies. Hardly any study in this field has looked sufficiently at the increasing private provision of technical services to armed forces, its reasons and possible implications. There is no study, which examines the full spectrum of privately provided technical services to the armed forces. This book intends to fill this gap in the literature.

Although military privatization and the increased role of contractors in the US military can be traced back several decades, it is a quite new phenomenon with respect to other Western armed forces. At present we are in the middle of an ongoing process in which the US military is far ahead of other armed forces. Many things can occur in this process, but it seems highly likely that it will continue. Considering the fact that ministries of defence around the world are very busy at the moment ordering enormously expensive high-tech weapons for high-intensity warfare, which are exactly those kinds of weapons that require most contractor support, the present trend of military privatization is bound to become even more pervasive.[6] The new approach to defence procurement, best symbolized by the *Joint Strike Fighter*, is to involve the industry in the whole life-cycle of defence equipment from the initial design to decommissioning. As a result a new and far more services-oriented defence sector is emerging, which will be described in this book.

A particular problem of the methodology used in this study was to define 'services' and especially 'military services'. The term 'services' itself is confusing and can be basically applied to anything, which does not belong to the primary (agriculture) or secondary (manufacturing) sectors of the economy (Chesbrough 2004). When researching services within a business branch, which has traditionally been associated with manufacturing, a dilemma arises. Where can a line between the traditional defence industry, which just manufactures weapons and the new defence and security services industry, which sells hardware but also related services, be

5 Services like consulting or training might not be naturally considered 'technical services', but depending on the context they can be highly technical. Military consulting could mean to develop a plan for weapons procurement, so that various different weapons systems and components fit together. This would be systems integration and engineering work, but would still appear as a consultancy contract. Similarly, military training can, and increasingly does, involve the use of complex computer-based simulations and simulators. The services provided are predominantly technical, as they relate to the equipment used.

6 A simple look at the biggest procurement projects in the US, the UK, France and Germany makes it pretty obvious that those countries have already committed themselves to extremely expensive force modernization projects. The equipment that has already been ordered is most useful for force projection for short high-intensity conflicts. This new high-tech approach to warfare is outlined in important strategy papers such as the American *Quadrennial Defense Review* (2001/2005), Britain's *Strategic Defence Review* (1998/2002),the French *Loi Programmation Militaire* 2003-2008 (2002) and the German *Defence Guidelines* (2003).

drawn? The solution is to avoid rigid company or industry classifications and to consider 'military services' simply as an extension of the arms trade. These services directly or indirectly relate to technology, which has been acquired by the armed forces. As 'high-tech' is driving the civilian services economy,[7] it is certainly driving the 'military services' economy as well. Technology breeds the need for technical services. The more complex the technology and procedures, the more services can be built around it. That is exactly what is happening in the defence business.

There is little doubt that services are becoming more important in the global economy and that this development is connected to the present process of globalization. Services account for more than 70 per cent to the value added in advanced economies and to more than 70 per cent of overall employment and growing (Cuadrado-Roura, Rubalcaba-Bermejo and Bryson 2002, 5). At the same time services are becoming more transnational and account for more than 21 per cent of the international trade (Cuadrado-Roura, Rubalcaba-Bermejo and Bryson 2002, 9). It has been argued that the ' 'globalization' of goods cannot be conceptualized without services and vice versa; both depend on each other and this dependency is at the centre of the process of globalization' (Cuadrado-Roura, Rubalcaba-Bermejo and Bryson 2002, 4). For this reason, it is necessary to see the present globalization of the defence and security industries in close connection to the globalization of military and security services.

Services have certain characteristics, which distinguish them from pure manufacturing activities and which shall be briefly outlined. First of all, services are in contrast to material goods intangible. They need to be negotiated between a provider and a customer. The value of services is hard to measure. For example, if it took a mechanic five hours to fix a car, has he done a good job, or would it have been possible to do the work in two hours at half of the costs? It gets more complicated still, if the customer has no clear idea of a desired outcome. This is often the case in the initial phases of an R&D project. This means that services often require a more long-standing relationship between providers and customers and a mutual understanding of each others capabilities and requirements. The customer is directly involved in the creation of services and is a co-producer (Chesbrough 2004). Services tend to tie provider and customer together, especially in the high-tech areas. On the other hand, contracting out services gives both sides more flexibility, as contractual arrangements can be terminated or renegotiated. A service provider can also work for many other clients, which could raise questions in terms of allegiance and could lead to conflicting interests. It is also a main characteristic of services that they often need to be provided in the proximity of the client and that they are consumed when they are produced, which means that they cannot be stored. This applies to services in general, but has many important consequences for military services in particular. In short, services are very different to normal procurement and the services revolution in the defence sector is rife for research with the analytical tools of a future 'services science'. Up to now the research on services focuses on commercial businesses such as finance, business consultancy, transportation, tourism, or telecommunications.

7 The computerization of society is probably the biggest single reason for the growth of the services sector since the 1960s. Even the big computer manufacturers such as IBM make half of their money by selling services rather than hardware (Lohr 2006, 4).

Very little research has been done in the area of governmental and military services. Hopefully, this book will be also a contribution to this emerging academic field.

The book consists of eight chapters. The first chapter discusses the evolution of the Western armed forces and their professionalization since the late eighteenth century. It gives a historical overview and sets the stage for the discussion in the following chapters. The chapter shows how technology has impacted on the way the armed forces are organized and how they operate. With the growing complexity of weapons and tactics, the military had to become more complex itself. The military has also become increasingly dependent on its societal environment, which has to meet its growing support needs. With the nuclear revolution of the 1940s and the information revolution that began in the 1970s, the armed forces have become increasingly dependent on outside expertise. As a result of the so-called *Revolution in Military Affairs* in the 1990s and the goal of transforming the military into smaller high-tech forces, military contractors have taken over a crucial role in bringing about this transformation.

The second chapter outlines the tremendous changes, which have occurred in the defence industry since the end of the Cold War, among them the shrinking defence sector, mergers, export orientation and globalization of the defence business. This will establish the connection between the increase of military outsourcing and the transformation of the defence industry after the Cold War. Finally, the importance and impact of civilian technologies are discussed and the chapter closes with an assessment of the future of the defence industry. It will be shown that the IT revolution has substantially increased the importance of dual-use and commercial technology in the defence sector and that this has opened up the defence market to newcomers specializing in IT and related services.

The third, fourth and fifth chapters all deal with the various technical services, which are already provided by private companies to many modern armed forces. These chapters give some background information on military technical services and describe these services. Many examples are discussed and some trends and problems are analysed. The third chapter looks at military research and IT related services, including systems integration and military communications. The fourth chapter focuses on what is usually summed up in the defence literature as 'modelling, simulation and wargaming'. Particularly important is the involvement of private companies in military training, planning, simulation and development of doctrine. Finally, the fifth chapter addresses the various operational support services like maintenance and upgrading, operating systems and focused logistics.

The sixth chapter summarizes the most important findings and looks at the connected issues and possible implications. For example, it discusses the problem of having contractors on the battlefield, the legal status of contractors and the issue of the continuous loss of competence of the armed forces as result of the outsourcing. The chapter further explores some other issues such as the question whether the outsourcing of military services is cost-effective and how military privatization might impact on the principle of sovereignty. It will be shown that many military, economic and political issues related to military privatization remain unsolved, while the process itself continues.

The seventh chapter looks at the phenomenon of military privatization from a system level perspective. The Western armed forces operate globally and maintain extensive as well as intensive links to foreign armed forces. At the same time, the armed forces depend increasingly on globally connected and globally operating private companies. The introduction of complex military equipment such as network-centric systems and the outsourcing of military services have many interesting implications for international security. Security relationships between nations are becoming more and more complex because of the impact of technology and the privatization of many military activities, which reduces chances for effective government control of military capabilities. The export of more complex equipment also usually implies that the exporting countries provide the support capacities for the equipment, which could be withdrawn at any time. This means that there is more uncertainty on both sides: the arms suppliers have less control over the degree to which technology is transferred, while the recipients cannot be sure, whether the support for imported systems can be secured in the long term. The growing complexity of security relations between states generally increases the uncertainty and instability of the international system. This will continue to make war a very risky business, especially for the nations with the technologically most advanced militaries.

The concluding chapter summarizes the main argument that technology is a driving force in the present military privatization. It is argued that technology has a life of its own leading to many unintended consequences. Therefore it remains dubious that the transformation of Western armed forces into smaller high-tech intervention forces can help much to safeguard international peace and security. It might as well cause an adverse outcome.

Chapter 1

Technology and Privatization

The military as a distinct social institution is as old as warfare itself. In the course of its history the military evolved as a social organization and became more complex. The development of new weapons led to new tactics and diversified the armed forces into different types of troops specializing in particular weapons. As more advanced weapons and more complex tactics could make a difference on the battlefield, it encouraged further innovation and diversification. In this process warfare became increasingly a business, as armed forces developed more complex needs in terms of weapons, equipment, training and logistics. This chapter attempts to trace the evolution of the military as a subunit of modern society with a focus on the interaction of technology and the development of the business aspect of war.

Evolution is a biological concept that explains the ability of an organism to acquire more complex features and to adapt to its environment. Evolution is not straightforward and has no final aim. Rather it can be interpreted as a blind mechanism that produces a tendency towards higher complexity (e.g. Dawkins 2000). A high degree of complexity, however, is in itself neither good, nor bad for the organism or system concerned, but brings with it its own set of characteristics and challenges. The German sociologist Niklas Luhmann has applied the concept of evolution to society in order to understand the development of social organizations and society at large (Luhmann 1995b). More recently military theorists have become interested in concepts like evolution and complexity and have used them for explaining military history or current military challenges (e.g. Gray 2003; Hammes 2004). The following section aims to explain how the modern military has become what it is now and to what extend technological complexity impacts on its organization and mode of operation.

From the Middle Ages to the Invention of the Military-Industrial Complex

Although change is a predominant feature of warfare, the process of innovation was generally very slow in most of its history. Innovation occurred only occasionally, but became much steadier from the late Middle Ages onwards. Technology was mainly important in naval warfare and for siege warfare, but overall there were only marginal technological changes from the campaigns of Alexander the Great to the campaigns of Ghengis Khan. Human strength and the strength of animals used in warfare, such as horses and oxen, had remained the main source of energy and military power (Keegan 1993). The European knight was the eclipse of the attempt of maximizing the effects of human and animal strength by combining high mobility and survivability to deliver shock. For centuries mounted knights dominated the

European battlefields and foot soldiers were usually no match for them. Knights sometimes fought dismounted, but during the times of Charlemagne and up to the twelfth century there was practically no infantry as such. However, when military forces learned to utilize more complex tactics, the dominance of the knights was eventually broken.

The battle of Crecy fought in 1346 is usually seen as an important turning point in military history. The English troops were outnumbered and largely on foot with little cavalry. Still they were able to defeat a force of mounted French knights two times as big, which was up to then unheard of. This was made possible by the effective use of the longbow, a weapon that requires a lot of skill, but is cheap to produce. The English archers were well trained and highly disciplined, while military discipline was not part of the military tradition of the European knight, who considered himself as a free man and a landowning warrior. This led to disaster for the French knights and proved that disciplined dismounted forces can be highly effective. As a result, in the late Middle Ages the small warfaring class of knights had to be augmented by large numbers of men on foot for their protection against enemy infantry in order to be effective. Knights became eventually so vulnerable to the new weapon of the crossbow that its use was outlawed in Europe in 1139. However, this could not stop the terminal decline of the medieval practice of warfare, which had begun long before the battle of Crecy. More foot soldiers could be fielded instead of knights, which meant that they needed to be equipped with crossbows, pikes and other weapons. According to William McNeill this increased demand for weapons led to the beginning of the commercialization of the European arms trade in the fourteenth century (McNeill 1984, 68, 69).

In Italy, at that time Europe's economic centre, warfare quickly became a private business and subject to market forces. Large numbers of soldiers could not easily be conscripted from own lands because this disrupted economic activity. Therefore they tended to be recruited from abroad and to be led by war entrepreneurs, who were hired by the Italian city states. These so-called condottieri, or contractors, were freelancing captains, who were contracted to raise a force and to lead and maintain it for a clearly specified time, or a particular campaign. While militias were composed of amateurs, the mercenaries were professionals, as they earned their living through war. This meant that mercenary forces were usually more effective. Unlike the image of mercenaries (mainly created by Guicciardini and Machiavelli), the condottieri and mercenary companies usually served their employers well and they often got long-term contracts, which attached them to their masters more permanently. However, problems with regard to discipline and fidelity did arise because of delayed and insufficient payment, which occurred frequently because of the heavy indebtedness of the Italian states (Mallet 1974, 39).

Mercenaries specialized in weapons like the crossbow or the pike and essentially fought for anybody who would pay them (Fuller 1998, 74). Their high costs meant that mercenary armies were generally small. However, mercenaries often formed the nucleus of larger and cheaper conscript armies (Keegan 1998, 8). Deborah Avant pointed out that the mercenary companies of the Middle Ages often provided a wide range of services, which included apart from combat advisory, training and logistics services (Avant 2001). During this period mercenaries became important in warfare because of their greater military skills and their mastery of new technology. This sometimes afforded them to become powerful political actors. Notable in this respect are the free companies of the fourteenth century that dominated Italian warfare for almost a century. The gunpowder revolution in the fifteenth and sixteenth centuries led to the development and widespread use of artillery, which further encouraged the process of specialization and professionalization of the military trade. Military organizations initially lacked the expertise of utilizing gunpowder weapons and had to rely on specialists for hire. Contractors were usually used to transport, maintain and operate artillery, as no one else was qualified doing it. They also kept the finer details of their craft secret and formed guilds to protect their interests and expand their influence. In the end they could not prevent armed forces from eventually developing the capability to utilize artillery themselves. Consequently the contractors were no longer needed and had only the choice to change their profession or to join the regular forces.

With the invention of the gunpowder and the continuous improvement of the musket and the artillery, wars became more and more expensive and required a much more organized effort (Fuller 1998, 93; Münkler, 2005, 51). For example, traditional fortifications became obsolete, as fortresses had to be equipped with angular walls and salient bastions with cannons in order to defend them against enemy artillery. Further, military success now required forces, which had to be composed of infantry, cavalry and artillery, and they had to be used in a combined manner. All forces had also to be properly equipped and required years of specialized training, as the correct handling of weapons and a high degree of discipline became crucial on the battlefield. Therefore forces could not simply be dissolved after the end of hostilities, but had to be permanently maintained and trained, which only the emerging big nation states could afford by raising taxes, while smaller states and warlords were forced out of the business of war. As a direct result of growing technological and tactical complexity, military force began to be organized differently in modern times.

The modern military can be described as a complex adaptive system which emerged together with the nation state in the middle of the seventeenth century and the new practice of maintaining standing armies. From then on the armed forces evolved independently in the sense that they became a self-reproducing, self-organizing and self-sustaining subsystem of society. One important step in this development was the rise of military professionalism since the late eighteenth century when the first military schools were established in the Western world. While it was previously possible for nobles and other upper-class people to simply buy a commission, it became increasingly uncommon and this practice was finally abolished in the first decades of the nineteenth century.[1] Being an officer required then training and ability. This recognized need for leadership training clearly indicates that the military leader needed more than training on the job, but also a theoretical understanding of warfare.

1 John Keegan points out that the purchase of a commission was abolished in France after the French Revolution of 1789 and in the United States in 1802, but continued to be possible in Britain as late as 1871. However, all major European powers established military academies for officer training in the early nineteenth century (see Keegan 1987, 177).

Technology played and plays an important role in the rise of military professionalism. The first military schools, most notably West Point, were established as engineering schools for artillery officers and army engineers. Van Creveld explains: 'With complex technology demanding high skills, and at the same time helping make available textbooks to teach those skills, military training was gradually turned into a formal affair comprising courses and examinations.' (van Creveld 1989, 145) At the same time, the pace of technological development increased and industrialization allowed equipping and fielding larger armies, which had increasingly to be led by highly trained professionals. However, mercenaries and other private actors in warfare did not completely disappear. They still played an important role in warfare up to the end of the nineteenth century, but mainly in the context of colonial wars outside of Europe.

While mercenarism was in decline because of the rise of national armed forces, the role of civilians in warfare shifted to the logistics aspects of war. The creation of huge conscript armies after the French Revolution increased the dependence of these armies on private contractors 'for the provision of food, clothing, shelter, transport, labor and general housekeeping' (Shrader 1999). As a result, contractors accompanied these forces in substantial numbers. The production and trade of weapons, however, was still largely controlled by the emerging nation states. In the period between 1650 and 1850 there was a great stability in the European system of arms production and trade, which was caused by the process of standardization of weapons used by the European armies (Krause 1992, 54, 55). Weapons were produced in state-run factories and there were little differences in terms of their sophistication across Europe. This only changed slowly through the gradual privatization of the arms trade. In times of war private arms manufacturers could provide the additional quantities of weapons and ammunition needed. But once established private arms manufacturers attempted to stay in business even after the end of wars by pitching new weapon designs to the armed forces, or by selling arms abroad (Kaldor 1983, 23–24). The development of a private arms industry and the commercial competition among the manufacturers therefore hastened the process of developing new and improved weapons.

In the late nineteenth century national armed forces entered a more intimate alliance with the weapons manufacturers to better control the process of innovation. The armed forces set up their own research laboratories and employed their own engineering experts and specialists. McNeill points out that 'Navy technicians set out to specify the desirable performance characteristics for a new gun, engine, or ship, and, in effect, challenged engineers to come up with appropriate designs. Invention thus became deliberate.' (McNeill 1984, 278) When the sale of advanced weapons by private companies to potential adversaries became an issue (in 1882 the British Armstrong shipyard sold a technologically advanced cruiser to Chile, which sparked a debate in Parliament on British naval security), the government wanted more control over the arms industry. In return, the industry got assurances from the government to cover investments in R&D. At that point the military-industrial complex was born. From then on also began the coevolution of the military and its scientific and industrial bases. The military stimulated innovation in weapons design and then had to become more professional in order to cope with the higher technical sophistication of its weaponry.

The Complexity Leap in the Twentieth Century

The most important single factor in driving the complexity of the military system is technology. Not only does this mean that the military has used more and more complex weapons, but it also had to develop more complex tactics and a more complex organization in order to support them. With the development of gunpowder weapons, the military had to trim its organization in a way to ensure supply of ammunition. With the development of petroleum fuelled vehicles the complexity of supporting a motorized force in the field has increased several-fold. This led to impressive gains in military effectiveness, but it also enormously increased the dependency of the military on its societal environment, which has to meet its increased support needs.

Managing the large mass armies created by the draft required new organizational capabilities in order to cope with the information necessary for effective management. The result was the creation of the modern general staff, which can be according to van Creveld understood as a data-processing unit. The size of these industrial age armies, which makes the task of command much more complex, required a general staff and the general staff created possibilities for further growth (Fuller 1998, 117; van Creveld 1989, 235). Still, the complexity of the internal and external support infrastructure for the military has increased tremendously in the twentieth century. This meant that the military had to significantly change itself. The military had to become more complex itself in order to deal with the increased environmental complexity and the increased complexity of its own support requirements. As van Creveld pointed out:

> there is no comparing a modern self-propelled gun with its tens of thousands of precision-made components, to the artillery of 50, or even 15, years ago. However, the complexity of individual pieces of hardware only forms a relatively small part of the problem. Other factors are the momentous variety of equipment used in modern war; the need to back up that equipment with hundreds of thousands, even millions, of different spare parts that often require different kinds of storage facilities and have different expected life spans; the need to train, organize, and cater to the needs of the many classes of specialized personnel who alone are capable of maintaining, repairing, and operating the equipment; and the task of merging both hardware and personnel into integrated teams, capable of surviving on the battlefield and of fulfilling their missions under the tremendous pressures it brings to bear. (van Creveld 1989, 235)

The first high-tech weapons, which already made the problems of employing sophisticated technology quite clear, were the modern battleships of the late nineteenth /early twentieth century. The new ship designs were so complex that they exceeded the competence of the naval officers who ordered them (McNeill 1984, 294). The ships were also so expensive that only few of them could be built. It additionally became obvious that these few battleships needed to be manned by relatively few professionals and that in naval warfare there was already no place for the masses of rudimentarily trained conscripts. So despite the fact that military expenditures in Europe almost doubled between 1900 and 1914, this increase of expenditure was not accompanied by an increase in manpower. The money went primarily into new complex weaponry, such as field guns and battleships (Kennedy 1984, 8). In the

First World War technology enabled the Great Powers to field unprecedented large armies because of easier mobilization, transportation and sustainment in the field. But it also began to show the irrelevance of mass mobilization. Mass mobilization led to four years of military stalemate and only the economic breakdown of the Central Powers sealed their fate. It was believed that the next war would be decided by smaller professional armies, which would be much more capable of utilizing new technology. Indeed the huge conscript armies of the Second World War were not so much needed because of their size, but because of the much increased complexity of the armed forces. Michael Howard explains this by comparing the differences in equipment between an infantry battalion of the First and Second World War. The infantry in the First World War had just two or three standard weapons. 'In the Second World War fighting units were highly diversified. The inventory of the simple infantry battalion contained not only rifles and grenades but two types of mortars, two kinds of machine gun, light tracked vehicles, anti-tank guns, hand-held anti-tank weapons, and several types of mine. The demands of armoured units were many times more complex; those of amphibious and airborne units more complex still.' (Howard 2001, 133)

Managing this diversity and complexity required great numbers of people for administration, catering, maintenance, transportation etc. Many people serving as conscripts in the mass armies of the Second World War were not directly involved in fighting, but were performing work very similar to their civilian occupation as administrators, cooks, mechanics and so on. Martin van Creveld has pointed out that the number of Military Occupation Specialties (MOS) has even increased substantially in absolute as well as relative terms since the Second World War. This added to organizational complexity as a result of a process of military specialization and diversification caused by a revolution in technology.[2] It became more and more difficult for the military to cope with its own complexity and the complexity of the technology it utilized, which led to a greater role of civilians in all aspects of warfare.

In the twentieth century contractors increasingly took over responsibilities to support equipment and weapons used by the armed forces. The introduction and use of more sophisticated weapons such as aircraft and tanks required the assistance of the manufacturers for repairing and maintaining them. This issue emerged during the Second World War when American manufacturers had to send technical personnel into the theatre of operations in order to ensure that their products performed well and to take care of any technical problems that arose under operational conditions. Charles Shrader notes that:

2 Van Creveld claims that 'Toward the end of World War II the 9,700,000 men enlisted in the U.S. Services were divided into 1,407 different Military Occupation Specialties (MOS's), the average number trained in each MOS being thus 6,894 … By 1963 the number of enlisted men in all four services was approximately 2,225,000 in 1559 MOS's, so that the average number of men in each MOS was down to 1,427 – and even this figure misrepresents the true situation because many of the 1963 MOS's, especially those connected with the rapidly growing field of electronic gear maintenance and operation, were really agglomerations of several different specialties.' (van Creveld 1985, 235-239).

> For the first time in World War II, the manufacturer's technical representative became a prominent feature in forward areas. The increased complexity of military aircraft, signal equipment, vehicles and other war material produced by American factories, the rapid introduction of new models, and frequent changes in operating and maintenance requirements made the 'tech rep' a welcome addition to forward airfields, depots and repair facilities. In some cases, tech reps were even to be found in the front lines seeking solutions to technical and operational problems regarding equipment supplied by their firms. (Shrader 1999)

During the Second World War American contractors also built and operated ordnance factories in North Africa and the Middle East, which were at these times war zones. After the Second World War the so-called contingency contracting became important, especially in the Korean and Vietnam Wars. This had political rather than technological reasons. As there were political limitations on the number of troops that the US could devote to these conflicts, civilians had to temporarily augment the shortages of military manpower in theatre. During the Vietnam War most contingency contractors were involved in military construction and logistics work related to base operations. For example, the construction conglomerate RMK-BRJ (Raymond International, Morris-Knudsen, Brown & Root, Jones) employed in 1967 over 51,000 people for building airstrips, digging channels for ships and constructing military bases all over South Vietnam (Bricdy 2004, 164). Even more extensive is the role of Kellogg Brown & Root (KBR) during the occupation of Iraq, as it is the single contingency logistics contractor for the US forces, housing, feeding and otherwise supporting 130,000 US troops on 45 sites. With respect to these contingency contractors the situation is quite comparable to the Vietnam era. However, some technological developments that began during the Second World War did substantially alter the role of civilians in warfare in other, more crucial aspects.

The Nuclear Revolution

The emerging Cold War and the nuclear and missile revolutions led to a further level of complexity. With the danger of nuclear war the military found itself again in an unprecedented situation with enormous uncertainties. The military had to prepare for a kind of technological war, which had never been fought before. Nuclear weapons threatened to make all conventional military capabilities and traditional soldiering irrelevant. The problem of nuclear war seemed to render military experience and tradition meaningless, as war would be a purely technical matter in which a few bombers and missiles in a short high-intensity war decided the fate of the nation. The most important questions of what kind of weapons and delivery systems were needed, or how they would be deployed and used, could not be figured out intuitively or by experience, but required a high degree of technical understanding. A decision-maker had not only to consider the technical capabilities of individual weapons systems, but also how they interact with each other. For example, the question whether a bomber should have higher payload, but shorter range and air-refueling capability, or have long range with less payload and no air-refuelling capability depends on a range of interconnected factors, which need to be taken into consideration: number of targets, number of bombers and other delivery systems, hardening of targets, air-refuelling capabilities, number of bombers expected to get through to the targets etc.

This kind of problem led to the rise of *systems analysis* and the civilian defence analyst, who claimed to have the answers to these questions. The judgements of the people from the Santa Monica based think tank RAND, who came up with systems analysis, used complex models and simulations, which were mathematical as well as intuitive and rivalled those of the combat experienced military officers. Sharon Ghamari-Tabrizi claims that 'Atomic weapons inaugurated a colossal shift in authority. They swallowed up the personal wisdom of senior officers rooted in combat experience in favor of intuitions arising from repeated trials of laboratory-staged simulations of future war.' (Ghamari-Tabrizi 2005, 48) However, the danger of nuclear war did not completely make conventional war obsolete and the superpowers still had to maintain large conventional forces. The Soviet Union needed large conventional forces in order to keep its political hold over Eastern Europe and the US to be able to counter a possible Soviet attack on Western Europe without being forced to rely on nuclear weapons only. The Vietnam War already showed the limited value of conscripts in the emerging form of high-tech war and the US could abandon the draft in 1973 without having to fear a decline of its military capabilities.

The Information Age

The first indications of some major changes in conventional warfare came with the first use of precision munitions during the Linebacker raids at the very end of the Vietnam War. Laser-guided bombs were able to destroy targets that had consistently evaded destruction before (Friedman and Friedman 1996, 237). The US and its allies added more and more conventional precision weapons to their arsenals, which had become deadly effective. During the same period new powerful space-based technical intelligence capabilities were developed and fielded. This includes the improved American Keyhole spysatellites with a digital film readout system, which could send satellite images back to the earth in near real-time, the Rhyolite satellites, which could intercept Soviet communications all over the world and also the monitoring system, which was established for tracking Soviet fleet movements. Military command centres and nuclear submarines were connected through space-based communications systems and the ARPANET, the predecessor of the Internet, was set up as a survivable communications infrastructure for fighting a nuclear war.

However, new command and control systems became also very important for conventional war. Some evidence of the increased effectiveness of conventional precision weapons combined with improved intelligence capabilities and command and control systems could be seen during the 1973 Yom Kippur War. Israel, which was supported by the US materially and with intelligence, managed to obliterate a numerically vastly superior Arab force consisting of thousands of tanks. Computers, information networks and 'smart' weapons were destined to transform warfare and the Soviets were the first to theorize about it (Cohen 1996, 39).

Termed *Military-Technological Revolution* (MTR) in the late 1970s by General Nikolai Ogarkov, it was based on the idea that information technology could revolutionize military operations. The Soviets spoke of reconnaissance-strike complexes, which refers to the integration of shooters and sensors for automated troop control (Krepinevich 2002, 6). This concept is similar to the American concept of network-centric warfare developed in the 1990s. Unable to exploit this MTR because of underdeveloped Soviet computer technology, the Soviets did not attempt to transform their military. However, the Americans, who put less emphasis on military theory than the Soviets, were developing new tactics and fielded new high-tech weapons in the 1970s and early 1980s as a response to the Soviet numerical superiority in conventional arms in Europe. This was called by the US Secretary of Defense Harold Brown 'offset-strategy' and aimed at gaining a decisive qualitative advantage over the Soviets (Sloan 2002, 25). Already in the 1970s there was the idea to develop the so-called *Assault Breaker* weapon system, which would stop the second and third waves of troops of a Soviet attack before they would reach the frontline. Although *Assault Breaker* was not successfully developed, it changed the US/NATO outlook on strategy in Europe and threatened to make the entire Soviet military strategy obsolete (C. Gray 2005, 107). The *AirLand Battle* and *Follow-On-Forces Attack* (FOFA) doctrines, which were some kind of extension of *Assault Breaker*, emphasized offensive operations as the optimal defence against a Soviet attack. Instead of trying to simply defend Western Europe on NATO's Eastern border, the new doctrines opted for deep conventional strikes into Warsaw Pact territory. In this concept of an 'extended battlefield' the enemy would be attacked in the full depth of the battlefield, which requires synchronized air and ground operations. *AirLand Battle* was finally tested in the 1991 Gulf War and it gave a glimpse to the possibilities a true MTR would offer (Krepinevich 2002, 9).

The American success during the 1991 Gulf War sparked a general debate in Western defence establishments about a *Revolution in Military Affairs* (RMA), which potentially lay shortly ahead. As a result, many states around the world became interested in military modernization programmes in order to realize the potentials of an RMA. Chris Demchak claims that there were at least 68 modernizing states in 1997. According to him, they all shared a similar vision of a future military based on the American concept of an RMA that emphasizes network-centricity, range, accuracy, speed and lethality (Demchak 1999, 186). It is certainly true that in some respects the new IT enabled high-tech weapons created unprecedented new powerful military capabilities, but they also caused many unforeseen problems.

'Rogue Outcomes' and 'Normal Accidents'

The emergence of smart precision weapons during the late Vietnam era has brought the complexity of military technology to a level, which makes it enormously difficult for the military to cope with it. The information age has created a rather paradoxical situation in which technology on the one hand increases the ease of operation of a weapons system, while tremendously complicating the required support for it. There is a study by Chris Demchak, which nicely illustrates this problem. He shows how the introduction of the M1 battle tank into the US Army in 1982 has affected its organization and what kind of detrimental and unexpected effects the use of high-technology has in warfare (Demchak 1992, 347).

In the early 1980s the M1 was to replace the older M60 tank and it incorporated some amazing new high-technology such as a stabilized high-powered main gun, which allows hitting targets while the tank drives 30mph. There were worries about the tank's operational performance, but these proved to be unjustified. Far more important, as Demchak points out, were the organizational consequences the introduction of the tank had. The tank itself was actually easier to operate than the M60 and it has a built-in test system, which can automatically diagnose errors. The training requirements for the crews and maintainers could be reduced and the Army managers subsequently cut back maintainers, test equipment and spare parts.

The big drawback of this system were its extraordinarily high maintenance and support needs because of its internal complexity, which was no longer accessible to the crews and maintainers alike, as the maintainers could only exchange what were for them 'black boxes'. First of all, it is very inefficient to exchange a complex part every time it fails, because this would require having a large stockpile of very expensive components. Secondly, such an approach increases the logistics burden several-fold in a military operation because there is never just one tank, which needs to be supported, but maybe hundreds with thousands of different components. Thirdly, the test equipment was also far less reliable than expected, which meant that the time for locating the error increased significantly adding more uncertainty to the military organization (Demchak 1991, 104).

The big problem with complex technical systems is that they contain lots of unknown errors or 'rogue sets', which cannot be predicted (only operational experience can be gained in the long term). The result was that these 'rogue sets' had to be accommodated by buying maintenance contracts from the manufacturers, by developing additional internal procedures and by making organizational changes. According to Demchak, the Army responded to this logistical challenge by increasing the number of people in an armoured division from 19,000 to 20,000 and the number of maintenance personnel from 1,311 to 1,562. The tank companies lost their internal maintenance capability, as maintenance had now to be organized in a more centralized fashion on the battalion level and the inventories of spare parts also increased by 34 per cent, while the operating costs were 41 per cent above the M60. Because front-line maintainers no longer had the skills to repair components, contractors had to step in, which meant that 'The use of contractors was integral to the introduction of the M1 into units.' (Demchak 1992, 347) The knowledge burden for the Army to support the M1 was so high that it could not support the tank with internal resources alone and had to secure external help.

There are lots of lessons that can be learned from this example, as it applies to most other high-tech systems as well. The initial aim of the Army of rationalization through the introduction of a high-tech system caused an adverse outcome. Supporting the more complex system required more people, more skills and more money. The Pentagon's ambition to build ever more complex weapons systems, which require fewer or even no operators, might only cause an indirect increase of required personnel and skills elsewhere. Charles Perrow, who has analysed the effects of high-tech systems, observed that 'Rationalisation of a system has replaced the operators; it is the province of designers and managers. This process has a long history. But rationalization allows more complexly interactive systems to be built, and the failure manager, the operator comes back in.' (Perrow 1999, 270) The error, or in Chris Demchak's parlance 'rogue set', becomes an inevitable result of the operation of a complex technical system. Thus the accident becomes a normal occurrence, or simply a 'normal accident'. The greater complexity of a technical system means also that the role of the operator is less important because the operator has little actual control over what is happening. Much more important are the roles of the designer, the manager and the trouble-shooter because they provide him with, or safeguard the capabilities the operator simply uses. In the case of an automated system the real operator *is* the trouble-shooter and not the person, who deploys the system. The soldier, who operates a complex weapons system, has effectively very little understanding of it. Understanding, however, is also not even necessary, as some newer high-tech weapons are actually easier to operate, which only reinforces the knowledge gap between designers and operators.

An obvious parallel are small computers, which were in 1980 still so difficult to use that only computer enthusiasts could use them. Today they are so easy to use that everyone can learn the basics in one hour. If a computer breaks down and the problem can not be solved by simply exchanging a defunct component, the expert is as indispensable as ever and as today's computers are much more complex, the skills and knowledge of the trouble-shooter needs to be proportionately higher.

In some cases it is not yet possible to get rid of the operator altogether, for example in civilian airliners or military aircraft. The required skills of the operators have therefore substantially increased because they need to be able to act as trouble-shooters as well. A pilot not only needs to know how to fly the aircraft, a task which has gotten incredibly easy thanks to automatic flight controls, but also needs the engineering knowledge to understand how the flight systems work. This enables the pilot to interfere effectively in case they fail. It is pretty clear that the average soldier cannot have the same level of expensive training as jet pilots and also not the engineering skills for understanding the technology that is employed. This means the system contractor, who has the engineering knowledge, is becoming indispensable and integral to the success of military operations. Peter Singer quotes an analyst, who argued: 'We're using the most advanced technology in the history of the world to wage wars and sometimes the people who built it are the only ones who know how to fix it.' (Singer 2003, 64) The soldier can only hope that the designer has done a good job, or if not, that a competent trouble-shooter is available for dealing with errors and 'normal accidents'. As a result, every time modern armies go to war they have to take private contractors with them, who have to be in the theatre of operations and are exposed to considerable risk, as stand-off weaponry and terrorism has made the distinction between a hot frontline and a safe rear virtually obsolete.

Not surprisingly, system contractors are sometimes deployed within military units and do essential work by keeping weapons systems operational in the field. Contractors accompany armoured units during combat operations, or serve on warships (Talmadge 2004). Occasionally they even operate these systems, while the soldiers watch (Robinson 2002). There is not only a growing army of systems contractors taking care of particular military equipment, but there is also a growing number of IT contractors, who set up and maintain computer networks in a theatre of operations and manage satellite communications. The role of private companies in providing technical services has therefore tremendously increased with the information revolution of the 1990s. This reality has not only been accepted by the military, but was also politically embraced in a general policy of privatization and outsourcing, which has tremendously increased the role of military contractors.

Towards a Postmodern Military

Considering the great problems and costs of military high-tech and its limited usefulness, one could ask the question: why was the idea of an RMA so readily accepted by Western military establishments? There are clearly some important non-technological reasons for this general tendency towards a high-tech military, which need to be considered. Social change has immensely contributed to the current obsession of modern armed forces with military high technology. A general change of values in society impacts largely on recruitment and on the context and manner in which military forces can be used.

With the decline of major wars and the increased importance of economic growth, the idea of a global civil society has become quite popular around the world. Ideals and values have changed and Western societies have been described as being post-military, post-materialistic, or simply post-modern. Individualism and personal self-fulfilment have become much more important than nationalism and sacrifice for abstract ideas. People with postmodern values put a high emphasis on enjoying a high living-quality, on doing meaningful work and being able to unfold all personal talents. Francis Fukuyama has argued that this signals a rather positive development and that the world is now headed towards democracy, market economy and peace (Fukuyama 1992).

How did the present situation develop? First of all, as there are no aggressive neighbour states left in Europe and North America against which the military has to defend the nation, there is the growing impression that the military is no longer important and that high military expenditures simply cannot be justified. There is also the impression that the military lacks a clear purpose and direction, as warfighting no longer appears to be the primary mission. This creates uncomfortable uncertainty for the overall military profession. The declining importance of the military for national defence is also accompanied with the declining prestige of the military. It is no longer regarded as an honour or even a duty to serve the country. People from the wealthier and more educated classes of Western societies increasingly evade military service. The armed forces are now more and more urged to recruit from the lower and middle strata of the society, who go there not because of conviction, but because of a lack of alternatives (Snider 2000, 10).

This clearly shows the general unattractiveness of military service. Most Western militaries experience great difficulties retaining and recruiting personnel. They are hardly able to maintain their current numerical strength. Recruitment shortfalls are one of the main reasons why all Western armed forces are increasingly becoming smaller. In fact the numbers armed forces personnel in the world has fallen from roughly 26 million in 1992 to about 20.5 million in 2002 and there was not a single Western country where armed forces personnel has increased in 2002-2003 (BICC 2004). Military service is unattractive because it generally means a lot of sacrifice in regard to individual freedom and also hard work for comparatively little payment. As the challenges for the armed forces have grown, the average work hours have gone up. With the substantial increase of the number of military operations in the context of the UN and NATO, soldiers can expect to be stationed overseas and to spend months or even years in boring and dangerous work places, which seriously complicates family life.

The problem for the armed forces is not only that fewer people volunteer for service, but also that they no longer get the best of a generation and often have to reduce their standards (J. Wilson 2005). This is a serious problem, as the job of a soldier becomes more complex and difficult and involves the handling of information age technology, which requires a higher educational standard and longer training. For this reason many Western states have already abandoned conscription or are in the process of abolishing it.[3] The new military missions simply require a higher degree of professionalism than can be achieved with conscripts, who only serve a relatively short time and who cannot be deployed in dangerous places for political reasons (Freedman 1998, 15).

The new professional information age militaries have to rely on highly motivated and talented volunteers with good technical skills, who are very difficult to find as such people tend to prefer civilian jobs which are better paid and more convenient. It is also often the case that particularly the most talented officers and NCOs leave the forces long before retirement age in order to start a second career in the business world (Townsend 2006). Often militaries cannot afford this continuous loss of talented and experienced personnel and if the trend continues, they might face significant drops in military effectiveness. Additionally, as Christopher Coker points out, there is a shortage of people, who have the necessary physical ability, as 'Westerners today are physically weaker than at any time in their history'. He adds that 'Even the U.S. Marine Corps has had to reduce its standards of physical fitness simply to make good a shortfall in numbers enlisted.' (Coker 2002, 157)

To fill the ranks armed forces have begun to recruit more women and people from ethnic minorities, but even this might not be sufficient for keeping force levels. Therefore militaries might be increasingly willing to recruit foreigners, who would be considered mercenaries, as they would serve mainly because of individual gains such as money or a better life. Steven Metz calls this option of recruiting foreigners 'one additional form of the privatization of security' (Metz 2000, 21). Restrictions for recruitment in regard to national, racial or ethnic origins no longer seem to make any sense. This situation will get even far worse in the future because of demographic developments in Western countries, which will drastically reduce the pool of potential recruits (Schindlmayr 2002, 85-91). As a result, there is little alternative for Western

3 Britain, Canada, and Australia abolished the draft in the 1960s, the US in 1973, more recently Belgium (1992), France (2001), Spain (2001), Portugal (2003) and Italy (2006). Other countries like Germany and Russia are clearly taking steps in that direction (Hau 2003).

forces than substituting technology for manpower by developing weapons systems which require fewer and fewer operators while promising to be more effective than current ones.

Another major social factor that reinforces the importance of technology is the changed context in which Western armed forces are used. Postmodern wars are fought without serious national interest. They are interventions far away from home and are justified on humanitarian grounds or on the basis of some other vague principles. Operation Restore Hope in Somalia was the first of a whole series of humanitarian interventions in the 1990s. In such interventions there are no vital national interests at stake for the nations, which take part in them. The goals and stakes are fairly limited and do, therefore, not justify a high price. For example, the US did not defend vital national interests in Somalia in the early 1990s. When the intervention eventually did cost the life of 18 American soldiers, it caused public disagreement as the costs of life (which were, in fact, quite moderate compared to other wars the US military fought) were still regarded as being too high. Consequently the US pulled out and from then on rarely committed ground forces (Shawcross 2001, 65-70).

Instead air strikes became the favourite means of projecting military power. This new strategy became obvious in the air campaigns over Iraq in 1998 (Desert Fox) and over Kosovo in 1999 (Deliberate Force). Christopher Coker points out that: 'Averaged out, President Clinton launched a cruise missile every three days of his term' (Coker 2001, 59). The advantage of missile and air attacks is that they promise zero casualties and can be aborted anytime, if necessary or when politically desired. It also helps to reduce enemy casualties, which are publicly considered to be unacceptable as well. Unfortunately, wars without casualties can hardly be decisive and as Kosovo showed, aerial bombing from high altitudes can be quite ineffective. Steven Metz suggests that this could mean the disappearance of the concept of decisive victory (Metz and Millen 2003, 13).

Because of the new predominance of stand-off precision weapons Edward Luttwak claims 'that U.S. ground forces are not available as instruments of U.S. foreign policy, except under very unusual conditions' (Luttwak 1996, 33-45). He calls this new trend of military intervention by air strikes post-heroic warfare, as heroes and heroism are no longer needed or even wanted in such wars. The risk of life is minimized and the average soldier expects the best protection from the perils of the job, which means that not only Western societies, but also their militaries are getting soft. Hardly anyone, who signs up for military service, expects having to die for the own country. In fact many more civilians working for NGOs were killed in action than UN soldiers in the 1990s (Lilly 2001, 184). According to Martin Cook, even all-volunteer forces can no longer be expected to engage in peacekeeping without sufficient force protection because of the moral contract between soldier and state (Cook 2004). They serve for the defence and the vital interest of their nation and might legitimately object to be exposed to unnecessary risks. Future humanitarian interventions can only be justified on the grounds of minimum risks for civilians and the soldiers entrusted with them (Shaw 2005, 79). This again means that the development and use of advanced military technology that can minimize dangers to civilians and soldiers (such as precision munitions and military robotics) becomes imperative and a precondition for the legitimate use of military force.

Conclusion

The chapter has attempted to trace the evolution of the Western military and to show the impact of growing technological complexity on military organization and tactics. Greater complexity made the military increasingly dependent on its societal environment, which has to provide the military with weapons, supplies and logistics services. Managing technological change and growing complexity required the military at times to rely substantially on external material support and expertise. The degree of the privatization of military activities was greatest in times of revolutionary change and smallest in times of great stability.

The historical overview began with the late Middle Ages, which saw tremendous change in the practice of warfare because of the decline of the hereditary military caste of European knights. The higher discipline of the infantry compared to knights allowed more complex tactics and additionally the infantry's weapons had a greater range. These features allowed foot soldiers to become more effective than mounted knights, who needed then to be protected with infantry against enemy infantry. This substantially increased the number of combatants fielded in campaigns and therefore also the demand of weapons. Arms production became commercialized and armies were increasingly composed of foreign mercenaries, who specialized in particular weapons like the pike or the crossbow. The gunpowder revolution that occurred in the fifteenth and sixteenth centuries made warfare much more expensive, as it required a more organized approach to war. In the end it forced mercenaries and warlords in Europe out of business.

In the time between the establishment of the Westphalian system in 1648 to the middle of the nineteenth century there was great stability in military affairs. There were few significant military inventions and there was a tightly state controlled system of arms production. This changed significantly in the late nineteenth century when the gradual privatization of arms production led to an increasing pace of military innovation. The military-industrial complex emerged and revolutionary change occurred several times in the course of the twentieth century. Most important are probably the nuclear revolution and the information revolution. The nuclear revolution caused a tremendous shift in military culture, as civilians came into a position to determine military strategy. The information revolution almost achieved a similar shift in military culture with respect to conventional warfare. The new information systems, military networks and computers chips that were incorporated in virtually every peace of military hardware required civilian experts for managing the new technology. Complex technical systems produce 'normal accidents' and this needs to be accommodated by having the ready support of the manufacturers. As a result, system contractors have become integral to the success of military operations.

Finally, the chapter tried to explain why Western armed forces became so obsessed with high-tech despite the fact that it causes operational difficulties and seems to be irrelevant for many forms of military conflict. It was argued that social change is a strong reinforcing mechanism of the technological revolution. Attitudes towards the military profession have changed, which negatively affects recruitment and retention. At the same time new military requirements and technology has made the citizen army largely composed of conscripts obsolete. As a result, Western armed forces are becoming smaller and smaller and the only hope they can have keeping similar levels of readiness and effectiveness as during the Cold War is through the utilization of military high-tech. Computers serve as substitute soldiers and can for political reasons be more easily exposed to danger than real soldiers. In other words, military high-tech provides part of the legitimacy for Western military intervention.

The next chapter will look at the emergence of a new military industry that was triggered by the end of the Cold War, but which has its roots clearly in the technological revolution that has occurred in the 1970s and 1980s.

Chapter 2

A New Military Industry

The defence industry has undergone significant restructuring since the end of the Cold War. As NATO's main threat has largely disappeared, the Western defence industry found itself without a strategic competitor and hence without a need for continuous innovation and rearmament. With the decrease of national defence budgets, procurement of new weapons plummeted as well. The defence industry quickly recovered after significant downsizing and major defence contractors on both sides of the Atlantic are producing good revenues again because of the American and (to lesser degree) European commitment to an RMA. A direct effect of the RMA is that it makes warfare more capital intensive, as high-tech weapons and particularly the necessary R&D connected to them, is much more expensive. This has led, according to Ann Markusen and Sean Costigan, to 'a more recent trend toward privatization of defense research, services, and depots', which is expected to continue (Markusen and Costigan 1999, 9). From a Western defence industry perspective, the RMA first and foremost means a different mode of doing business. Instead of manufacturing great numbers of sophisticated platforms, which are no longer needed in high quantities because of the doctrinal shift from platform-centric to network-centric operations (Dombrowski, Gholz and Ross 2003), the new emphasis is on software and services rather than on manufacturing. This might help the industry to remain healthy in the long term and will further change the defence business, which is no longer the sole domain of the traditional weapons manufacturers. The remaining defence companies will increasingly move away from a manufacturing-based business concept to a more service-based one. It will also become much more globally interconnected than it already is and will have to change even more radically in order to survive.

Defence Industry Trends

The trends discussed below refer particularly to the defence industry in the United States and Western Europe, but some of them are also global trends. The US and Western Europe (UK, France, Germany, and Italy) account for 75 per cent of global arms production (Bitzinger 2003, 6)[1] and are therefore most important for understanding global defence industry trends. There are some significant differences in the development of the defence sector between the US and Europe. But generally speaking, the major trends like defence industry downsizing, company mergers,

1 Within the EU defence production is concentrated in the UK, France and Germany and to a lesser degree in Italy, Spain and Sweden. Those six countries account for '90 per cent of Europe's defence industrial capabilities, 85 per cent of EU total defence spending and 98 per cent of all R&D expenditure.' (See European Union 2005, 12).

industry privatization and a new export orientation are simply more pronounced in the US than anywhere else, which can be mainly explained with the market size and political unity of the US compared to Europe.

Shrinking Defence Sector

During the Cold War the West had overestimated Soviet military spending and capabilities. In fact the Soviet Union had devoted a much greater amount of its GDP to defence expenditures than the US and other Western countries, but its backward economy could hardly support this. The high costs of keeping up militarily with the West contributed to the collapse of Soviet socialism (see e.g. Miller 1998, 382; Norquist 2002, 1-41). With the end of the Soviet Union, Russia significantly reduced military expenditures, troop levels and major weapons systems, and is still hardly capable of paying its soldiers and maintaining its most advanced weapons systems, including its nuclear forces (Zaloga 2000, 12-14). The bloody war in Chechnya, the rotting Black Sea fleet and the Kursk submarine accident in summer 2000 underline the general impression that Russia will not be able to pose a serious military threat in the foreseeable future. Russia's defence expenditures now amount to less than a fifth of the US (Latham 2003, 9). With the Soviet threat gone and only a few potential 'rogue states' left, the US and other Western countries cut back defence budgets in the early 1990s and the lowest point was reached during the mid-1990s (SIPRI 2004). The US defence budget of 2000 represented only three per cent of the GDP compared to over six per cent in 1985 (Carter and White 1998, 138). In Western Europe the cuts in defence expenditures were slightly less pronounced (UK MoD 2005, 25).

This was felt badly in the defence industry, which had to cope in the early 1990s with cancelled or significantly reduced orders and stretched or postponed forces modernization projects. Although a large part of US and European military equipment was about to reach the end of its planned lifetime at the late 1990s, Western governments hesitated at that time to commit themselves to necessary modernization projects. As a result, military procurement spending dropped 27 per cent in the US and 38 per cent in EU countries between 1988 and 1996 (SIPRI). Particularly the military aircraft industry, which produces the most expensive pieces of military equipment, came under pressure because of the cuts in the defence budget. The prices for military aircraft had also consistently increased since the First World War, so that one observer noted that if this trend continued, the entire US defence budget could purchase only a single tactical aircraft in 2054 (Callum 1998, 105). Nevertheless, the situation became almost obscene with the costs of delays and the enormous overcapacities of defence manufacturers. An illustrative example is the *Eurofighter Typhoon*, which was first conceived two decades ago (mid 1980s) and which is already ten years out of date, although according to procurement planning, the last planes are to be delivered to the British and German forces in 2015. With overall costs of £19.7 billion, the Eurofighter is one of the most expensive current defence projects and it seems that it is only because of the high amount of money already invested that this project, which has lost its military purpose, is continued (Feuchtwanger 2004, 63-68). An extreme case is the B-2 *Spirit* bomber of which the

Pentagon originally intended to buy about 132, but then decided to only purchase a much lower number. In the end only 21 B-2 bombers were built for $1.6 billion a copy. The production line was finally closed in 1998 (Correll 1998), making it the most expensive single tactical weapons system ever, with rather dubious military value.[2]

With the unit prices going up, the purchased numbers have gone down (Fallows 2002, 62-74). Another example of this mechanism is the F/A-22 Raptor, which certainly has outstanding performance characteristics (supercruise, extreme manoeuvrability, stealth invisibility), but it comes at an insanely high price. This post-Cold War project was in the early 1990s already so expensive that none of the then five big US military aircraft manufacturers (Lockheed, Boeing, General Dynamics, Northrop, and McDonnell Douglas) could complete it alone. Lockheed partnered with Boeing and General Dynamics to share the development costs and enormous risks (Rich and Janos 2003, 350). Although Lockheed won the contest against Northrop, the project itself appears to be no longer very profitable, as the Air Force has now decided to purchase a much lower number of the plane (less than 200 instead of 800) (Crock 2005, 66), bringing the costs per unit to almost $260 million (Weiner 2004) and making it the most expensive fighter jet ever to have entered production line.[3] Congress has now put a cost ceiling of $37.6 billion on the procurement of the F/A-22, which will certainly be difficult to meet.

Similar is the picture for many other expensive defence items, in particular for sophisticated major weapons platforms. Some of them have been cancelled after a decade or more of development (Crusader artillery, Comanche gunship, Sea Shadow stealth ship), or have significantly changed over time (SDI to National Missile Defense to Theater Ballistic Missile Defense). With rapidly falling procurement spending in the first half of the 1990s and growing R&D costs, the defence industry had to undergo substantial restructuring.

Defence Mergers

The overall situation pointed towards industry consolidation and radical downsizing. Worldwide employment in the defence industry has dropped sharply from 16 million to 11,5 million between 1985 and 1994 (Stanley 1998, 35-66) and was in 2002 only 7.67 million (BICC Survey 2004). Overall, employment in the industry fell even faster and was more drastic than in the armed forces.[4] Western governments have therefore encouraged conversion and the merger of large defence contractors in

---

2 Theo Farrell claims that the B-2 bomber was procured despite the absence of a strategic requirement. The US did not need a long-range nuclear bomber after the end of the Cold War and its enormously high costs make its adaption to conventional bombing missions unreasonable (Farrell 1997, 61-66).

3 The US DoD has finally (after 14 years of development) approved the full-rate production of the F/A-22 on 19 April 2005.

4 The armed forces of the world have shrunk 25 per cent since the end of the Cold War, while the defence industry has shrunk more than 50 per cent.

order to deal with the reduced global demand for weapons.[5] Attempts of defence companies to move into the commercial business were sometimes not successful (Gregory 1993, 84) and in most cases they tended to stay at least partially in the armament business (Gregory 1993, 57; Markusen 1998, 121-146).

A series of mergers since the early 1990s has changed the defence industry tremendously. The trend of mass mergers of defence giants was much more pronounced in the US, as the European industry was already quite concentrated within the European states and different national procurement priorities have traditionally caused the higher degree of fragmentation (SIPRI 2004). In the US, mergers created such giant corporations like Lockheed-Martin (Lockheed merged with Martin-Marietta, GE Aerospace, Loral and the aircraft branch of General Dynamics), which makes $35 billion in revenues a year. The second biggest defence company is now Boeing, which had merged with Rockwell and McDonnell Douglas and which is more successful in the military business than in selling civilian airliners.[6] Third ranks Northrop-Grumman, a merger of Northrop, Westinghouse, Grumman and, most recently, the satellite producer TRW (in 2002). Northrop-Grumman offers a similarly vast spectrum of military products and services as Lockheed Martin and does everything from aircraft, nuclear submarines to developing and operating satellites. Fourth is Raytheon, the manufacturer of the Tomahawk cruise missiles, which has merged with GM Hughes, E-Systems and Texas Instruments. Raytheon specializes largely in electronics, network-centric systems (C4I), missiles (Hawk, Patriot, AMRAAM) and space systems (geospatial intelligence and spaceborne command and control) and the company reported in December 2003 $18 billion in sales. These four companies secured contracts, which amounted to about half of the overall DoD spending on research and procurement in 2003 ($58 billion out of $122 billion dollars).[7] In effect, the Pentagon cannot do much without them.

Table 2.5 shows that in several defence product areas there are now only two or three manufacturers left, meaning that the Pentagon has effectively very little choice in terms of manufacturers when it orders a next generation weapons system.

5 US Defense Secretary Cohen invited the representatives of the defence and aircraft industries to the 'last supper' in the White House in 1993, where he announced that the fat years are over and that the industry had now to consolidate itself by mergers and downsizing.

6 Boeing produces parts of the F/A-22, the F-15 and F/A-18 fighter jets, the Apache helicopter, the JDAM and is involved in missile defence research (Airborne Laser), as well as the sustainment of ICBMs.

7 According to the Center for Public Integrity database, Lockheed Martin received $22 billion, Boeing $17 billion, Northrop Grumman 11.2 billion, and Raytheon $7.6 billion for Pentagon contracts in 2003. The US DoD spent $54 billion on RDE&T and $68 billion on procurement in the same year. See Center of Public Integrity, www.publicintegrity.org/pns/list.aspx?act=top, accessed 06/06/2006.

Table 2.1 US Contractor Presence for Selected Military Platforms (1990-2000)

Platform	Companies 1990	Companies 2000
Fixed-wing Aircraft	8	3
Launch Vehicles	6	3
Rotorcraft	4	3
Satellites	8	6
Strategic Missiles	3	2
Submarines	2	2
Surface Ships	8	3
Tactical Missiles	13	3
Tactical Wheeled Vehicles	6	3
Tracked Combat Vehicles	3	2

Source: P.J. Dombrowski e.a. (2002), *Military Transformation and the Defense Industry after Next*, Newport, Naval War College, p. 22.

The picture is quite similar in Europe, where there is also very little domestic competition in the national defence markets left. For example, BAE SYSTEMS dominates the British defence industry, as it is now the largest employer in the defence sector and is involved in virtually every major British/international (European) defence project, including the Joint Strike Fighter, the Future Transport Aircraft, the Type 45 Air Attack Destroyer and the Future Carrier (Feuchtwanger 2004, 15 and 75). With the acquisition of Alvis Vickers in 2004 and United Defense in 2005 BAE SYSTEMS can manufacture the whole range of military equipment from aircraft, land vehicles to ships.

In France the Thales corporate group, which was created in 1997 through mergers and acquisitions of some of the biggest French defence companies (Alcatel, Thomson-CSF, Aerospatiale, Dassault Aviation and Lagardere), has effectively become the country's main prime contractor in the aerospace business.

In Germany, Daimler-Chrysler is now the dominant defence contractor, a development that began with DASA (Deutsche Aerospace Agency) formed in 1989 (Markusen and Costigan, 349) and continued with its partial ownership of EADS[8] (European Aeronautic and Defence Systems), a huge European (German/French) corporate group focusing on the aerospace and defence markets, which was formed in 2000. EADS includes Airbus (civilian airliners), Airbus Military Aircraft (A400M), Eurocopter (Tiger helicopter), EADS Space Transportation (launcher systems) and EADS Astrium (Galileo, Skynet 5, Helios 2B, SPOT 5). Originally the French and the Germans owned equal parts of EADS, but with the recent decision of Daimler Chrysler to reduce its shares, shareholding is now in imbalance (Daimler Chrysler 2006). There are still some other significant players in the German defence market left, most importantly Thyssen (Rheinmetall AG, Blohm and Voss) and Krauss-Maffei-

8 Daimler Chrysler holds about 30 per cent of EADS, the French SOGEADE holds another 30 per cent, the Spanish SEPI about 5.5 per cent and about 34 per cent are publicly traded.

Wegmann (Leopard 2 tanks), Atlas Elektronik (maritime and electronics systems, acquired by BAE SYSTEMS), but those companies have begun to either refocus their activities on civilian products or are trying hard to survive on arms exports.[9]

There is also only one big player left in the Italian defence industry: Finmeccanica, a consortium of companies, which includes Fiat-Iveco (armoured vehicles), Agusta Westland Ltd. (helicopters), AerMacchi (trainer aircraft), Galileo Avionica (command and control systems), Alenia Aeronautics (transport aircraft), Alenia Spazio and Telespazio (space systems).

Table 2.2 shows the world's top ten defence companies. More than half of them are American and the remaining are all European companies. The only exception is the state-owned Russian defence export company Rosoboronexport.

Table 2.2 The Top 10 Defense Firms Worldwide

Company	2001 Defense Revenue (Millions of US $)
Lockheed Martin (US)	22,502.0
Boeing (US)	19,000.0
BAE SYSTEMS (U.K.)	14,491.8
Raytheon (US)	11,969.0
Northrop Grumman (US)	9,337.5
General Dynamics (US)	7,784.0
Thales Group (French)	5,581.8
EADS, NV (European)	5,504.6
TRW (US)	5,200.0
Rosoboronexport (Russian)	4,200.0

Source: *Defense News*, 11-17 November, 2002.

New Export Orientation

Decreased defence spending at home led to a new focus on export, which became possible through reduced export restrictions, even for more modern systems like the General Dynamics F-16, which has been sold to 24 countries altogether. During the Cold War arms exports were allowed for mainly political or ideological reasons (bolstering allies and gaining influence), but now economic considerations have become more important (Gold 1999, 254).

Arms exporting has the advantage that it makes expensive defence products cheaper in two ways: first it produces revenues not only for the manufacturers but for the exporting states as well, and secondly, it allows the production of larger numbers, thus reducing the price per copy for domestic procurement. There are of course some constraints, most of them having to do with concern over technology

9 German military spending has declined between 2002 and 2004 and is significantly below that of the UK and France, while Germany is still among the world's top five arms exporters (SIPRI 2005, 318, 354 and 453).

transfers and the falling international demand for weapons. For example, the US is still trying to stay technologically far ahead of its European and Eastern competitors and is therefore very reluctant to share technology even with her closest allies (Taylor 1999, 66). The latest weapons systems are therefore generally still off-limits for export.[10] However, what has become quite common since the late 1980s are so-called military offsets, which are defined by the US General Accounting Office as 'a range of industrial and commercial compensations provided to foreign governments and firms as inducements or conditions for the purchase of US military goods and services' (US GAO 1997). For purchasing military equipment states increasingly expect some sort of compensation, which can take the form of 'mandatory co-production, licensed production, subcontractor production, technology transfer, countertrade, and foreign investment' (Jones 2001, 108). There are also other forms of direct or indirect compensation, which can look quite bizarre from a business point of view, as the exporting nations sometimes end up paying for the privilege of their companies to sell weapons.

Often exports are simply seen as a cheap way of getting rid of obsolete surplus weapons, which have high maintenance costs and are in the present security environment no longer needed. Ken Silverstein claims that 'As new weapons systems roll off production lines, the Pentagon can't even use, let alone store, its entire current arsenal.' (Silverstein 2000, X) Western countries have in general tremendous overcapacities of major weapons systems and with the overall shrinking armed forces the problem is exacerbated. The US is giving away great numbers of surplus weapons for free.[11] This contributed to the overall falling global demand for weapons. Although many Third World countries at the receiving end were happy to acquire even very outdated Western equipment so cheaply, the overall demand for weapons has shrunk in most parts of the world, except the richer and more conflict prone areas in the Middle East and South/ East Asia.[12] The international market for sophisticated military equipment is tight

10 However, some states with an advanced defence industry (first-tier producers) develop high-quality defence items primarily for export purposes because customers can demand the best in the tight global arms market. This is particularly true for Russia and other former Soviet states. In some cases the US has exported advanced weapons systems before they have entered service in the US armed forces (e.g. the Longbow Apache). It is also true that the F/A-22 is unsuitable for export simply because no nation in the world other than the US can afford to purchase a $200 million a copy fighter jet (US DoD 1999, 23; Wilson 2005, 89).

11 Lora Lumpe notes that: 'Since 1990, the United States has transferred approximately $7 billion worth of military equipment, including 3,900 heavy tanks and 500 ground-attack jets, primarily to developing countries. Large quantities of small arms and light weapons have also been exported. Giving away excess weapons is believed to be cheaper than destroying or storing them, and weapon transfers are seen as an easy way to gain favor. In 1990-94, 80 per cent of all excess transfers were giveaways; in 1995, half were.' (Lumpe 1996).

12 According to the SIPRI Yearbook 2005, military spending increased in the Middle East 40 per cent and in South/East Asia 20 per cent over the period 1995 to 2004 with increases in 2004 of 3.1 per cent and 2.7 per cent respectively. This and the fact that these two region have a limited defence industrial capacity makes them the most profitable markets for arms exports (SIPRI 2005, 310 and 420).

and export-orientation might not actually help to finance new expensive weapons projects. One illustrative example is the new French fighter jet *Rafale*, which was intended as a replacement for the ageing *Mirage 2000* and which has cost France overall $9 billion in R&D. In order to make the project profitable Dassault would have to sell 800 to 950 aircraft, for which neither a big enough domestic, nor international market exists (Callum 1998, 110).

Similar, if not worse, is the situation with respect to exporting main battle tanks. As the scenario of fighting major tank battles against advancing Soviet armoured forces racing through Western Europe has faded from history, the utility and future of heavy armour is very much in doubt. Few countries are still ordering tanks and the competition for these orders is tough (as can be seen in Table 2.3). Table 2.4 shows the world's biggest arms exporting countries and the value of these exports.

Table 2.3 Current Competition for Tanks

Recipient Country	Number of Tanks	Country of Competing Company
Turkey	1,000	Fr, Ger, Pak, Rus, Ukr, US, It, Israel
Greece	500	Fr, Ger, Rus, Ukr, UK, US
Saudi Arabia	450	Fr, UK, US

Source: *Jane's Defence Weekly*, 5 July, 2000, p. 3

Table 2.4 Top 10 Arms Exporters

Supplier Nation	Sales in millions US $ from 1999-2003
USA	29,599
Russia	26,198
France	6,372
Germany	5,240
UK	4,204
Ukraine	2,195
Italy	1,648
China	1,528
Netherlands	1,226
Canada	1,184

Source: SIPRI (2004), *Yearbook 2004*, International Arms Transfers.

Defence Industry Globalization

The defence industry has become much more globalized since the end of the Cold War, especially with respect to Anglo-American defence companies. While the Cold War period was marked by largely nationally controlled defence establishments with a strong emphasis on self-sufficiency in arms procurement for strategic and political reasons, this has changed significantly. Many defence companies have been privatized in the 1980s and 1990s and or have lost most of the protection and subventions they used to enjoy, which means that they 'have become much more directly subject to market forces' (Lovering 2000, 163). This provided the defence industries with lots of new challenges, as well as opportunities. Less state control and more openness for international cooperation allowed the big players in the defence business to expand internationally and to foster joint ventures with foreign companies.

As demand is falling and the costs for new weapons projects have substantially increased, there is the economic urge to share the costs and the risks of defence projects through international joint ventures. It has been pointed out that 'No European government can afford alone the costs of developing major items of equipment for the next century.' (MacDonald 1999, 15) The reason for the rising costs of weapons projects are the largely increased R&D costs because of the trend towards complex high-tech systems. This is particularly true for the aircraft industry. For example, the *Joint Strike Fighter* is the first American fighter plane, which will be built in cooperation with non-US companies. Another example is the Medium Extended Air Defense System (MEADS), which is a joint venture of the US, Germany and Italy (or better Lockheed Martin, EADS and Alenia Marconi Systems).[13] There are numerous joint ventures between European defence companies, for example BAE SYSTEMS works together with the French and the Italians on missile projects through the jointly owned missile producer MBDA. BAE SYSTEMS has also a joint venture with the Swedish Saab Military Aircraft for its fourth generation fighter *Gripen*. The 'French' Thales is now based in ten different countries and has started a long-term joint venture with the American Raytheon by founding Thales Raytheon Systems in 2001 to work on air defence and command and control centres, ground based air surveillance, and weapons locating radars (Lorell e.a. 2002, 171). According to Richard Bitzinger, 'Nearly all European armaments production is now collaborative, and increasingly rooted in international enterprises such as Eurocopter (a Franco-German helicopter joint venture), MBDA (an Anglo-French-Italian missile company) and Astrium (a British-French-German-Italian satellite builder).' (Bitzinger 2003, 68) Also notable is a much higher level of transatlantic defence industry cooperation and one important factor in this is that US companies have started buying British defence companies and vice versa, or opened subsidiaries on the other side of the Atlantic. While other European defence companies have problems accessing the American market, already up to ten per cent of the US defence industry is owned by British companies.

The defence business has also become globalized because of international markets and investment, which led to international acquisitions and mergers of companies in the defence market. Again, BAE SYSTEMS is a prime example. Since 2000 BAE SYSTEMS has acquired several US defence companies, among them Lockheed Martin Control Systems, Watkins Johnson, Condor Pacific, and United Defense. Lockheed Martin has also increased its presence in Europe by creating Lockheed Martin UK Ltd. in 1999, which has already won many lucrative contracts

13 This controversial defence project has been recently (April 2005) approved by the German parliament, which means that it will now enter the design and development stage.

from the British MoD, including the Soothsayer Electronic Warfare Programme, the Combined Arms Tactical Trainer (CATT) armoured warfare simulator and also works on network-centric solutions for the British forces. General Dynamics UK is another example.

A quite interesting development is that international investment groups have also become very influential in the global arms business (Lovering 2000, 167). For example, the holding company Carlyle Group holds shares of defence companies on both sides of the Atlantic, among them United Defense Ltd. and Qinetiq. While Carlyle has recently reduced its defence holdings, other big investment groups have followed its example (Ismail 2004). The times when defence companies were considered national assets and were protected against international investments and take-overs are gone. An observer therefore noted that defence companies, 'especially those operating in the UK, are becoming more international in character, regardless of their home base' (Latham 2003, 69).

There is also a growing technology gap between first-tier weapons producing states[14] (US and Western Europe) and second-tier states like the newly industrialized countries in East Asia, Israel, Brazil, South Africa, Sweden and Russia, which jeopardises the survival of these indigenous industries. Their domestic markets are not big enough to support them and exposed to the tough competition of the global post-Cold War arms market, they are hardly competitive enough with the big players. First-tier suppliers have already begun to acquire large numbers of second-tier companies, which might suggest that second-tier companies can only survive as subsidiaries or junior partners of global defence firms like Lockheed Martin or BAE SYSTEMS (Bitzinger 2003, 70-74). It therefore makes sense for second-tier producers to focus on market niches, which in the long term means that a global division of labour is emerging, where certain countries and companies specialize in particular fields of the defence business such as missiles/munitions, aircraft, naval vessels, helicopters, satellites/space systems and armoured vehicles (Battilega e.a. 2000). The other side of the coin is that a rather small number of prime contractors now depend on a myriad of subcontractors from all over the world. As about 70 per cent of the value of a complex weapons systems is derived from subsystems the dependency on global supply chains is growing (Hayward 2000, 120).

A major weapon system might still be assembled in the country where the prime contractor is based, but most of its components are likely to come from abroad. The general globalization of high-tech industries, of which the defence industry is only one part, makes it much harder for countries to keep control over militarily relevant technologies and products, as different components are often developed and manufactured in several other countries.

14 Keith Krause has analytically divided the global arms market into first-tier, second-tier and third-tier producers. First-tier producers are those which can 'innovate at the technological frontier,' second-tier suppliers 'produce at the technological frontier and adapt them to specific market needs', and third-tier suppliers can only 'copy and reproduce existing technologies' (Krause 1992, 31).

Defence Business Post 9/11

Post- 9/11 US defence spending has substantially increased after a period of reduced expenditures during the 1990s. It is now back to Cold War levels with a $379 billion US DoD budget (2003)[15] and worldwide military expenditures of $956 billion in 2003 (SIPRI).[16] Andrew Bacevich claims that present US defence spending is 12 per cent above the Cold War average (adjusted to inflation). According to long range Pentagon plans, 'by 2009 its budget will exceed the Cold War average by 23 per cent' (Bacevich 2005, 17). The 9/11 terrorist attacks resulted in George W. Bush's declaration of a 'War on Terror' and caused an expectation of a more prosperous climate for the defence industry. The shares of major US defence contractors were the only ones that went up immediately after the New York stock market reopened following 9/11. The 2003 war in Iraq led to another, but smaller defence boom, because only relatively few weapons and munitions needed to be replaced (Cappachio 2003). Still, the War on Terror and in particular the Iraq War caused a world military spending increase of 11 per cent in 2003, which highlights the fact that the US alone accounts now for almost half of worldwide military spending. Also in some European countries, most importantly Britain and France, defence expenditures have significantly increased after 9/11 (UK MoD 2005, 25).

The other, and in the long run more important, reason for the increased spending is that the idea of an armed forces transformation finally got off the ground and the US forces now again spend more on research and procurement. In the long run the industry faces the problem that the US (and other Western) forces have a great amount of surplus weapons and are not likely to order more of them as long as they can be upgraded. The higher costs of cutting edge defence projects and the general switch from a platform-based to a network-centric approach to warfare will in the long term mean that weapons manufacturers will sell smaller and smaller numbers. Some of the currently most expensive procurement projects (both US and internationally) are the JSF, the F/A-22, and the F-18 E/F, which could cost altogether more than $350 billions. This in effect means that there is with such a big commitment in the long term not much money left for other expensive big ticket items. For example, the Boeing/Sikorsky RAH-66 Comanche and the Crusader artillery system have been cancelled.

Worse is the situation in Europe, where governments have far less money left for procurement because of a handful of megaprojects like the Typhoon, the A400M, the

---

15 This does not include expenditures for ongoing military operations in Iraq and Afghanistan and also not expenditures on nuclear weapons, which are hidden in the budget of the Department of Energy.

16 To be fair, global military expenditures are, after taking into account inflation, just about two thirds of the peak reached during the years of the Reagan build-up in 1987, but this only represents purely military expenditures and not all expenditures related to security altogether. Considering the changed nature of warfare in which crime and war increasingly overlaps, additional substantial cost increases since the end of the Cold War need to be added, e.g. the $100 billion dollars annually spent on Homeland Security (Elhefnawy 2004,170).

Meteor (BVRAAM) and MEADS.[17] It can be assumed that Western governments will continue to invest in some major defence projects, but are also unlikely to buy any item in large numbers. In fact procurement spending of the European NATO states has decreased five per cent in the period of 2000 and 2003 (SIPRI). It is also generally hoped that with the development of a network enabled capability platforms can become simpler, smaller and cheaper, so that 'complexity is to reside in the web rather than the node' (Dombrowski, Gholz and Ross 2002, 7). Overall, the Western armed forces will spend less money on actual procurement and put more money into upgrading and military research in order to stay technologically at the cutting-edge. This is getting more and more difficult, however, because of the growing importance of civilian technologies.

Civilian Technologies

Military research has since the end of the First World War a leading role for the development of civilian technology. Especially after the Second World War military research used to be far ahead of civilian research in some militarily relevant fields such as jet engines, computers, or rocket technology, which it pioneered (see e.g. Hambling 2005). Although states could and still can devote enormous resources to research and development for keeping their forces at the technological frontier, they have already lost the technological race in many respects to commercial companies, which have altogether a research budget available several times higher than even the wealthiest state. More and more money is needed to stay technologically ahead, as technical systems have become increasingly complex, which means that R&D for any single product or system has also become much more expensive.

Ashton B. Carter points out that, although the proportion of US research spending to the other G-7 members had remained constant since 1980 (1:1 or $240 billion together), the actual proportion of what the US DoD contributed to overall research spending has dropped tremendously. The share of US DoD research spending amounts now only to one-twelfth of the total compared to the one-sixth it amounted in 1980 (Carter, Lettre and Smith, 134). Although US military research spending is much higher than in any other country and still amounts to about one third of overall US federal government spending on research, it is only a small fraction of what private companies globally spend on research. According to Mike Macedonia, private sector R&D spending on IT accounts to 75 per cent of the US total and drives computing innovation (Macedonia 2000, 126). While the private sector is under constant pressure to increase the market share, the military sector lacks the market compulsion. As defence companies see fewer returns from their R&D spending through actual procurement contracts because of the relatively small numbers procured by armed forces or cancelled orders, they subsequently spend less on R&D compared to other big private companies (Carter, Lettre and Smith, 139). In Britain BAE SYSTEMS only ranks number three of the UK companies spending most on R&D and not by a

17 The Typhoon is Europe's new fighter jet (Eurofighter); the A400M is a military transport aircraft; the Meteor is a beyond visible range air-to-air missile; and MEADS is a missile defence system for medium range missiles.

narrow margin. Glaxo SmithKline spent in 2002-2003 more than three times as much on R&D (2,900 million) as BAE SYSTEMS (899 million) and AstraZeneca more than twice as much (1,900 million) (Feuchtwanger 2004, 49).

Another reason why militaries are often no longer at the technological cutting edge is that technology innovation cycles have shortened, while the average lifetime of military hardware is 15 to 20 years (with possible upgrades). Add another ten years for the bureaucratic procurement process then it becomes clear that from the initiation of a procurement project to the end of the lifetime of a weapons systems can extend more than 30 years.[18] This would mean that the military would have to plan its equipment 30 years ahead – an almost impossible task, not only because of the increasing pace of technological change, but also because of the change of military missions and requirements. The overall bureaucratic process behind military procurement policy has proved indeed to be far too slow for reacting to the increasingly rapid change.

This problem became obvious with the growing role of computers both in the military and the private sector in the 1970s. Manuel DeLanda explains:

> While in 1964 the military represented 90 per cent of the market, its share toward the end of the '70s was only 10 per cent. Part of the reason for this decline was a set of bureaucratic regulations called 'Milspecs,' a set of specifications and tests that did not keep up with the speed at which the chip was evolving, and thus became an obstacle to new technology for weapons systems. With the chip's density of components and speed of operation doubling every year since its birth in 1960, the internal screening procedures built into the military's procurement system simply could not keep up. (DeLanda 1991, 152)

The bureaucratic process of defining military specifications, which outline all characteristics any military item shall have, was far too slow for the pace of technological development and also far too expensive.[19] As a result new military technology is far more likely to derive from civilian research and applications. According to Martin Libicki, 'commercial information technology will continue to advance at a rapid clip', which means that 'With every year, more and more technology comes from the commercial side. Even before the Cold War ended, the leading role of the defense acquisition had begun to fade. Military electronics started lagging behind commercial electronics and could only hope to stay current through spin-ons of commercial technologies.' (Libicki

---

18 The German Air Force has used the F-4 Phantom for over 40 years (!) and only now with the delivery of the Eurofighter it is phased out. This is no exceptional case. The US Air Force has some planes in its inventory with a service life of 50 years (the remarkable B-52 bomber) and might be in service till the middle of the century. Same applies to land forces. The M1 Abrams tank is around since 1978 and there are no concrete plans to replace it. Military ships tend to be the oldest equipment in a country's arsenal. For example the British Invincible class carriers are approaching a service life of 40 years now and will not be replaced before 2010.

19 According to data from Motorola from the mid 1980s, the development time of a commercial semiconductor was 1 to 12 months, which compares to 17 to 51 months for a military semiconductor. The latter was also about ten times more expensive (Gregory 1993, 51).

1994) Commercial high-tech companies have now the potential to enter the defence business and compete with traditional defence companies.

A possibility to do so is offered by the growing importance of dual-use products. Many pieces of equipment, which the military needs, can be bought off-the-shelf from non-defence private companies. They can be used for civilian and for military purposes by simply making a few modifications. For example, a helicopter can be used to transport injured people to hospital or it can be turned into a 'gunship' by adding machine guns and rockets. As many high-tech products, like electronics, are inherently dual-use and derive from the civilian sector, it makes sense for the military to buy them off-the-shelf, instead of trying to develop something on their own. Defence companies, for example, develop specialized software for the military, but for general applications widespread commercial software is usually much cheaper and better than anything which can be developed in-house (Carter, Lettre and Smith 2001, 136). The obvious downside of dual-use items is of course that the military cannot get customized products and has to adapt products, which are already on the market. These might not have certain desired performance characteristics such as hardening against radiation and shock-resistance (Gregory 1993, 53).

The civilian high-tech industries have certain advantages. They can profit from economics of scale, as they can mass-produce products and then sell them in large numbers on the world market. The more is produced, the cheaper the product can be sold and the cheaper it is, the easier it is to sell it in large numbers. Traditional defence manufacturers follow a different logic because they depend almost completely on government orders. They are usually making their profits by charging high prices for a limited number of performance maximizing weapons systems, which means that they cannot benefit from economics of scale in the same way as the civilian industries can (Latham 2000, 163).

Many technologies, which are used in civilian products, are important for modern weapons systems as well, in particular electronics. The same electronics, which can be found in PCs, which control car engines, or are in many other smart appliances, are essential for making the most modern weapons systems work. For this reason weapons manufacturers can profit from the cheap prices for commercial electronics and integrate them into new weapons systems. This way old weapons can be upgraded with newest (commercially available) electronics and software and can become much more effective through increased precision, or through synergy effects by linking them to other systems. According to the Forbes magazine, the defence industry has adjusted to the new realities and tries to exploit cheap commercial prices for high-tech components by leaking technologies to them. The defence companies can then buy them to mass production prices and insert them into any military equipment (Huber 2003).

As a result, new weapons systems can be produced much cheaper, but only at the price that they are already outdated when they enter service. On the other hand, because of falling electronics prices, smart weapons have become as common and disposable as bullets. It has been pointed out that in the recent Iraq War some 70 to 80 per cent had 'some sort of smart-weapon capabilities, as opposed to just 30 per cent of weapons in the first Gulf War' (Wait 2003, 8). In the end, hardware is much less important than the software that makes it work. As long as older and proven weapons systems can be upgraded with the newest electronics and software, they will remain effective and can be used far into the twenty-first century without any need to replace them quickly.[20]

With the spread of dual-use technology and products it also becomes much easier for private contractors to support the military, as the necessary skills can be found in commercial industries as well (Manker 2004, 15). Some analysts such as Wayne E. Lee and Allan R. Millet have attributed the increased role and presence of civilian contractors on the battlefield primarily to the increased reliance of the military on civilian technology (Bredemeier 2003). Dual-use equipment has already opened up the military contracting business to companies and individual contractors outside the traditional defence industries. The Pentagon has now even a directive that mandates the procurement of commercial products, if alternatives to military products exist, or the procurement of dual-use instead of military-specific products whenever possible.[21] Other Western ministries of defence have followed suit.

Privatization and Outsourcing

Privatization has become a global trend because of the general shift in economic policy that began in the early 1980s in the US and in some European countries like Britain, France and Germany after the numerous economic crises of the 1970s. Ronald Reagan and Margaret Thatcher attempted most vigorously to reduce the government share of the economy in order to allow market forces to regulate the economy, instead of having continuous state interference, which distorts the market and makes it less efficient. The overall aim of neoliberal economic policy is to have as little government as possible, so that the economy can regulate itself. The state has to become leaner by cutting its expenditures and by also reducing the state's tax income. As a result, many state-owned companies were sold and many state monopolies such as mail, telecommunications and railways were given up, either by turning national companies into stock companies and/or by allowing private competition.

The trend towards outsourcing began at the end of the 1980s, when big companies like IBM realized that they were not efficient and therefore not competitive in the world markets compared to smaller, much more flexible players. Giants like IBM had become too big, too bureaucratic and less innovative. With the growth of globalization and the emergence of vibrant markets in South and East Asia, Western multinational companies suddenly faced serious competition, in particular in the high-tech areas. Such companies had to go through a phase of radical downsizing in

---

20 For example, the US Air Force even considered keeping the B-52 in service until 2045, which would be a service life extending over almost a century. Cost pressures could mean that very little futuristic military equipment will actually be fielded, especially as there is no urgent need for them. This opinion is shared by many defence analysts, most importantly Michael O'Hanlon, who has argued that it does not make sense for the Pentagon to spend billions on replacing proven equipment just in order to pursue a far too ambitious RMA (O'Hanlon 2000).

21 See DoD Directive 5000.1 from May 12, 2003.

the late 1980s/early 1990s and many workers and employees lost their jobs because of a decline in their market share and overall profitability (Lippitz, O'Keefe and White 2000, 168). On the other hand, this economic restructuring created new jobs, as the big companies had to buy in services from smaller companies, which they were no longer prepared to do themselves. Such a practice is of great advantage for the big companies, as they are now less constrained by employment laws, do not have to pay pension benefits and can contract workforce more cheaply when it is needed and save money when it is not (Mowshowitz 2000, 118). Another driving force behind outsourcing is that many high-tech products require such sophisticated and expensive production techniques that even the biggest companies do not have all necessary capabilities of their own and therefore rely on a myriad of subcontractors.

During the 1990s 'outsourcing' has become a major trend in all business branches. Peter Singer states that 'Global outsourcing expenditures topped $ 1 trillion by 2001' and that: 'Of the 300 largest international companies, 93 per cent outsourced some function.' (Singer 2003, 68) Not only big companies have used this strategy, but also governments have begun to entrust private companies all kind of functions, which were once the exclusive domain of the state.

Although the terms 'privatization' and 'outsourcing' are usually used synonymously, there are important differences, as pointed out by the lobbying group *Business Executives for National Security* (BENS) (Taibl 1997). Privatization means the transfer of assets or ownership from the public sector to the private sector, for example by selling state owned companies to private companies, or by transforming them into publicly traded stock companies. Outsourcing means that certain non-core functions, which used to be done in-house, are delegated to external organizations. Only states can privatize, while outsourcing can done by both companies and states. Sometimes governments combine both: the transfer of the ownership of facilities and other assets to private companies, which take over the management of these assets. For example, when a military depot is sold to a private company, which charges the government for the management of this depot, then privatization and outsourcing are blurred. However, if the government eliminates a certain in-house capability, for example the catering for troops in the field, and completely hands over this responsibility to private companies, then this is a clear case of outsourcing. More difficult is the case of defence manufacturers, which provide personnel to the military for services connected to training, maintenance and operation of new products they sold to the military. Technically speaking, it is neither privatization, nor outsourcing, as it does not imply the transfer of public assets and also not the elimination of certain capabilities, which previously existed. Rather it can be seen as the result of long-term strategic procurement decisions that implied the greater responsibility of defence companies for their products.

The Growth of the Services Sector

The defence industry in many respects is now undergoing a development, which commercial industries already underwent some 20 years ago. From a broader perspective it can be said that the most advanced industrial states had to go through a painful process of economic restructuring in the early to mid-1980s, which meant that the manufacturing sector of the economy has shrunk, while some new jobs were created in the services sector. According to Jeremy Rifkin, a long-term trend can be identified: in the 1950s, 33 per cent of American employees worked in the industry, in the 1960s 30 per cent and in 1980s 20 per cent and it dropped further to 17 per cent in 1995, with a further decrease to be expected (Rifkin 1995, 8). It is therefore estimated that services now account to about 80 per cent of the US gross domestic product (GDP) and about 60 per cent to 80 per cent of the GDP in other advanced economies (Chesbrough 2005, 43).

The general trend is that production is becoming more and more automated and requires fewer and fewer workers, while there is still a potential for growth in employment in the services sector, particularly for knowledge and IT-related jobs. New advanced production techniques like CAD and robotics make actual production of various different products with the same machinery much easier and faster. The manufacturing process is reduced to manipulating symbols on computer screens, no matter whether the product is steel, cars, or computer chips. This also means that the production techniques of civilian and defence products have become very similar. Often defence companies manufacture commercial goods in the same facilities they use for military products (Kelley and Watkins 1998, 251-280). Once a new product has been developed, it can often be rapidly produced and sold on the market in large quantities. At the same time it is more difficult to use rationalization in the services sector because automation would be in many cases too expensive, too complicated (e.g. work that requires creativity) or inappropriate (e.g. nursing). Therefore, overall employment in the manufacturing sector of the economy has fallen, while employment in the services sector has increased.

A similar development has occurred in the defence sector, where defence companies have now refocused their business to a greater service orientation. Instead of just manufacturing defence products and handing them over to the military, defence companies are now offering full support packages for their products, which means doing repairs, maintenance and upgrading as well, in addition to installation of systems and the training of the military personnel on their products (Manker 2004, 15). Generally speaking, the more complex a weapons system, the more maintenance is required to keep it operational and combat ready and also the more likely it is that the necessary expertise for maintaining it, is not found within the military. Military contractors now often do the maintenance work, which used to be done by the military itself and also offer 'life-cycle support' and management. As a result, the biggest growth within the industry has already materialized in profits made by maintenance of weapons systems, which coincided with the increasing willingness of governments to outsource these military related services. *The Economist* observed in an article on the Iraq War that less than 20 per cent of the 2003 US defence budget was spent on the procurement of new weapons, while the vast majority of costs related to 'operations and maintenance' (*Economist* 2003).

In the same article *The Economist* stressed that IT related services have become crucial 'to support an increasingly digital war'. The real winners of this restructuring of the defence sector are therefore companies like Halliburton, which is the biggest logistics provider for the US forces, but also big IT and communications companies, which have moved into the government/defence sector like Computer Sciences Corp. (CSC) by acquiring DynCorp, or L-3 Communications, which was founded in 1997 and is now a multi-billion defence contractor. As computer chips are now present in virtually every piece of military equipment, software is needed to make these weapons systems work. Lockheed-Martin, for example, writes millions of lines of software code a year and employs more software engineers than Microsoft.[22] According to a Pentagon study, 'software is becoming a more dominant, if not the most dominant, portion of a system's acquisition' (US DoD 2000, 11). This can be easily seen in respect to military aircraft where up to 80 per cent of the functionality requires software. A modern fighter jet packed with electronics such as the F/A-22 would simply fall out of the sky, if the flight computer fails even for a few seconds.

The distinction between manufacturing companies and services companies is increasingly becoming unpractical because there tends to be a lot of overlap. Hardly any defence company concentrates only on manufacturing and most companies offer goods and services, while the proportion of services related to goods is growing. For example, BAE SYSTEMS derives more than 50 per cent of its revenues from services and other defence companies aim at increasing their sales of services. There are also companies in the defence sector, which only offer services, but they are the exception rather than the rule. Table 2.5 shows that some of the biggest defence firms already produce a substantial amount of their revenues through services.

Table 2.5 Percentage Services of Selected Defence Firms

Company	Overall Revenues	Defence Related Revenues	Percentage Services
Lockheed Martin	$37.2 bn (2005)	$30 bn (70%)	15.3%
Northrop Grumman	$30.7 bn (2005) $26.4 bn (2003)	NA	32%
BAE SYSTEMS	£15.4 bn (2005)	£12.2 bn (79%)	+50%
Raytheon	$22 bn (2005)	$13.6 bn (62%)	50% (estimate)
EADS	€34.2 bn (2005)	€7.7 bn (23%)	NA
SAIC	$7.2 bn (2005) $5.9 bn (2003)	NA $2.6 bn (44%)	100%

Source: Lockheed Martin (2005), Annual Report; Northrop Grumman (2005), Annual Report; Raytheon (2005), Annual Report; EADS (2005, Annual Report; SAIC (2005), Annual Report.

22 See Lockheed Martin media 'IT: We are Lockheed-Martin', www.lockheedmartin.com.

The Revolution in (Military) Business Affairs

The services revolution in the defence sector is part of an attempt to completely redefine the relationship between the government and the defence industry in order to make better use of defence money. There were widespread complaints that the military bureaucracies are too inefficient, prone to cost-overruns and far too slow to cope with the rapid technological change. One particular concern was and is the so-called 'tail-to-tooth' ratio, which had become quite unfavourable (BENS 1997). Throughout the Cold War the ratio of combat capability to support was 50:50, now it looks more like 30:70 (Pages 1998, 382). More and more enablers are needed to support fewer and fewer warfighters.[23] In order to cope with today's challenges the Western militaries would have to accept the ideas of continuous improvement, innovation, competition and constant performance monitoring. It has been suggested that the military should concentrate on its 'core competencies' and leave everything else to the private sector (US DoD 1997, Section VIII). This basically means for the military not only to change fundamentally its relationship with the private sector, but also to change itself.

Some Western armed forces began to introduce modern management techniques in the 1990s and the idea that they should be run on a business model has become widely accepted in the Western defence establishments. The Pentagon has looked at commercial businesses for guiding its own privatization and outsourcing policies in order to reduce costs and increase performance (US DoD 1996). Former Undersecretary of Defense for Acquisition and Technology Jacques Gansler has called for a *Revolution in Business Affairs* (RBA), as this might be the only way to make the RMA affordable. According to him, the DoD has to change fundamentally the way it does business, which means a greater emphasis on cost-effectiveness and a greater involvement of the private sector in maintaining current equipment and in the overall transformation of the armed forces. Gansler suggested among other things base closures, the rapid reduction of military and civilian personnel, the 'drastic improvement of cycle times', the competitive outsourcing 'of all but inherently government functions' and an acquisition reform, which reduces bureaucracy caused by complex accounting and auditing systems (Gansler 1998, 30–33).

The overall goal is to have more flexible acquisition strategies so that the defence industry can quickly respond to the militaries' needs. Services and goods shall be provided 'on demand' and shall be delivered 'just in time' without the need for the militaries to keep expensive capabilities and spare parts permanently available. The Pentagon's Office of Management and Budget has now a guideline mandating 'that

23 Admiral Owens argues that there is too much redundancy in the US armed forces in regard to support functions. According to him 'overall for every U.S. "combat" soldier or Marine, there are about 12 "support" personnel. About 15 men or women back up the combat actions of a single sailor. For every fighter-bomber pilot in the US Air Force there are 32 other active-duty support personnel … Today 473,595 men and women in the US Army work in positions that support the 60,000 troops that are specifically equipped and trained for actual combat. At the same time 363,449 active-duty Air Force personnel support the remaining 16,000 who actually fly or navigate combat aircraft or operate their weapons.' (Owens and Offley 2000, 45).

the Government [shall] obtain commercially available goods and services from the private sector when it makes economic sense to do so' (Blizzard 2004).

Not surprisingly, the response from the defence industry to the idea of a RBA has been quite positive and the lobbying group BENS has tried to justify a greater involvement of the private sector in national security by finding evidence for successful privatization and superiority of the private sector in regard to efficiency. Also the Pentagon seems to be happy with the achieved savings. Stephen Blizzard has pointed out that: 'The GAO estimates that the average civilian support employee costs about $15K less than a comparably graded military person. The Air Force estimates that it has saved a $500M annually through privatization. DoD-wide cost savings were projected to be between $7B and $12B annually by fiscal year 2002.' (Blizzard 2004) The idea of a RBA, however, is not simply all about saving money and the present discussion therefore appears to be quite flawed. To understand these changes in the defence business it is necessary to look at the defence industry more closely in order to see why and how it is about to re-invent itself.

Where the Future Is

The revenues connected to software and services have substantially increased in the defence industry as a result of the RMA and tighter budgets for procurement. According to SIPRI, services accounted to 85 per cent of all contracts awarded by the DoD in 2003 (SIPRI 2005, 389). One American watchdog, the *Center for Public Integrity*, estimates that the Pentagon spends about half of its budget on procuring goods and services from private contractors and that 56 per cent of this sum was spent on services in 2003, meaning that revenues from services already exceed those from traditional procurement (Mackinson 2004). The big defence contractors have begun to set up their own technical services subsidiaries and to buy IT, logistics and security companies, as this is where they believe the future for the defence industry lies (SIPRI 2005, 389).

For example, Northrop-Grumman bought TRW Systems, which is now Northrop Grumman Mission Systems and which 'provides solutions for the entire program life cycle - from system architectures to the development and sustainment of mission-critical systems' including missile systems, command, control and intelligence and technical and management services, employing 18,000 people in 300 locations around the world.[24] This deal included BDM International, an IT company which is strong in the defence and government business and which was acquired previously by TRW. Vinnell, which provides technical services to the Saudi Air Force and also trained the new Iraqi army, is now part of Northrop Grumman as well. The technology company L-3 communications with a strong dependency on government/defence contracts and revenues exceeding $6.8 billion (2004) has bought MPRI, the biggest and most famous military consulting company. MPRI holds training and consulting contracts with the US military and other armed forces world wide and currently assists in setting up the

24 See Northrop Grumman Mission Systems website, www.ms.northropgrumman.com/home/main.html, accessed 28/02/06.

US Army Space and Missile Command (L-3 Communications 2004, 13). Computer Sciences Corp. (CSC), which claims to be the world's third largest provider for IT services, has also expanded its government business by acquiring DynCorp, which writes software and provides other technical services to US government agencies, but also does intelligence and security service work in Colombia, Afghanistan, Iraq, and Russia. Lockheed Martin also negotiated a deal to acquire Titan Corp., an IT company specializing in C4ISR solutions, but backed off in summer 2004 following the Abu Graihb scandal in which Titan employees were implicated in torturing Iraqi PoWs. Titan has now been acquired by L-3 Communications. Lockheed Martin had already merged with the IT company Loral in 1995 and has since then acquired lots of other IT companies. This short list could be easily continued. The overall picture is that all big defence companies have begun in the 1990s to substantially expand their business into the IT and services sector.

A consequence of the expansion into IT and other high-tech business is the growing diversification of companies in the defence business and their activities. Some prime contractors could successfully reduce their defence dependence by moving into the commercial business, but commercial companies moving into and expanding their defence business is a much more common pattern. An observer therefore noted that the 'So called 'hard-core' warfighting industry corporations all but disappeared.' (Friedman 2002) However, there are many civilian industries supplying the military and are ranking high on the military contractor list, although their military business might be quite small compared to their commercial business. It has been pointed out that: 'In the 1996 Defense News Top 100 List, for example, AT&T is listed as the company least reliant on defense, with only 0.5 per cent of its revenue derived from defense contracts. Yet it ranked number 73 in total defense contracts. Similarly, General Electric derived only 3.1 per cent of its revenue in defense, while ranking 21 on the total revenue list.' (Chu and Waxman 1998, 35-44) Many products and services purchased by the military are dual-use and were often not specifically developed for military purposes. This makes it quite difficult to determine which company can be considered a defence company and which one not. As a result, the defence industry is losing its distinct identity and this has implications for the way we perceive a certain company. The boundaries between defence and commercial sectors have become very fluid. It more and more appears that the once strong and distinct defence industry is increasingly becoming part of a global commercial industry with only a few megaprimes left, which produce little themselves and mainly manage a myriad of suppliers and subcontractors on which they depend almost completely.

Conclusion

This chapter has shown how the restructuring of the defence industry because of tighter defence budgets and falling demand has led to a far greater services orientation of the industry. The end of the Cold War forced the defence industry to reinvent itself and also created new business opportunities, as governments were willing to find innovative ways for meeting their defence needs.

The policy of privatization that began in the early 1980s was continued in the 1990s and also substantially expanded. When outsourcing became a major trend in the business world in the early 1990s, governments adopted this practice and began to outsource many government functions and services in the sectors of government administration, health, security, transportation, defence and many more. At the same time technology promised new capabilities and increased military effectiveness. The US wanted to take advantage of the RMA in the face of reduced military spending. This led the US DoD to call for a Revolution in Business Affairs to make the RMA affordable. The main idea of the RBA is that the armed forces will focus on their 'core competencies', which is 'warfighting', and leave everything else to the private sector.

Other Western governments were also interested in cutting defence expenditures and in keeping a sufficiently big defence industrial base. The shrinkage of the demand forced the industry to downsize and it also created defence giants like Lockheed Martin. The industry also increased its international orientation through exports and international collaboration and investment. As a result, the defence industry has followed other industries and is now already partially globalized. The 9/11 attacks have created a more prosperous climate for the defence industry and Western defence budgets have increased since then. Probably more important than this brief defence boom for the defence industry is in the long run the increased commitment of many Western governments to the idea of a transformation.

In the information age and era of globalization, governments have lost their lead also in the area of technologies relevant to defence to the private sector. Most innovation, especially when it comes to IT, occurs in the private sector and not so much in government sponsored research projects. IT products such as electronics or software are essentially dual-use products, which often derive from commercial applications and can be adopted for military use. The information revolution has therefore opened the defence business to non-defence companies like software firms or communications providers.

It has also led to the trend in the defence sector to move away from manufacturing and to focus instead on services. The big defence companies are now acquiring frantically logistics, software and security companies in order to prepare themselves for a future in which the money for procurement projects will be scarcer. The defence sector follows the general trend in advanced economies, which are already largely service economies. The boundaries of the defence sector have become more fluid and there are fewer defence specialized companies left. Some survivors have become mega-primes and are able to dominate the global defence market.

The following three chapters look at defence privatization and outsourcing initiatives presently underway in several NATO countries. About half of the examples refer to the United States and the other half of the examples refer to other NATO countries with a strong focus on the United Kingdom. This reflects the distribution of global military expenditure, as well as the fact that the US and the UK have pioneered military privatization. Additionally, most empirical information and experience that is available on this phenomenon derives from just these two countries. There is also a necessary focus on NATO countries for three reasons:

1) Their armed forces are similarly organized and equipped. With the American RMA underway and the need for the other NATO countries to operate together with US forces in coalition operations, they will have to develop their capabilities in a similar fashion and will therefore encounter similar issues and problems as in the case of the US.

2) NATO countries have the most advanced military forces in the world and the issues raised by ever more complex technology apply to them most. Russia, China and India might be catching up in the future, but this will take them at least decades, while Japan, which belongs to the technologically most advanced nations in the world still shows little interest in transforming its economic strength into military prowess.

3) There is much more information easily accessible on the armed forces of NATO countries than on any other armed forces. This is a purely practical consideration, but it is a reasonable assumption that the problems and issues related to the use of high-tech by the armed forces are universal and will be experienced by any armed forces which pursue the transformation to high-tech warfare.

Chapter 3

Research and IT-Related Services

There are, of course, many possibilities to divide the various technical services into chapters and subchapters. Finding a logical division of service types is difficult, as there is no generally accepted categorization of technical services and as nearly all of them have in some way or the other to do with computers. Necessarily there is quite a bit of overlap. In the following three smaller chapters various military technical services, which are performed by private companies will be explained in respect to their historical development, their nature, their importance or relevance and with the help of examples. As the field of military technical services is huge and also not very transparent, it is only possible to give a very general overview. It is therefore not within the scope of this book to give a complete catalogue of all technical services performed by private companies and all contracts awarded. These empirical chapters are rather an attempt to sketch the role and relevance of technical service contractors in information age warfare. This will highlight the intimate connection between technological change and the urge to transfer more responsibility to the private sector, which is a development that has now been going on for more than 50 years and runs parallel to the use and spread of computers.

The overall empirical section is divided into three smaller chapters: one chapter on research and IT-related services, which is subdivided into military research, systems integration and military IT. These different services are sufficiently similar to go into one section and also a similar set of companies and organizations are involved in them. A second chapter will be on modelling, simulation, and wargaming, which is subdivided into weapons development and testing, military training, and wargaming. This section specifically deals with the contribution of computer simulation and modelling for developing, management, training, and planning purposes. A third chapter will deal with operational support services and is subdivided into maintenance, logistics management, and the operation of systems. These services are more closely connected to military hardware and are closer to the operations side.

Military Research

Military research is a very wide area if looked at from a historical perspective. It is about using science to discover technologies relevant for developing new weapons (not necessarily the weapons themselves). Usually research and development are used together in one phrase (R&D), or even combined with testing and evaluation (RDT&E). Although these activities are inextricably linked together, they are still quite different activities. It is therefore reasonable to distinguish military research, which is about creating the scientific base for a later military application (basic research), from the actual development of particular weapons systems (and all the testing and evaluation activities which relate to it), which will be dealt with in the following chapter (4).

During the Cold War military research became a huge systematic and highly bureaucratized process, which drew upon a large infrastructure. It consists of national research labs, which set the general research priorities; the research labs of the individual services, which specified their particular needs; the research labs maintained by the defence industry, which developed new weapons according to national and service requirements; government and private think tanks, which formulated strategy and also guided military research; and finally universities, which also worked on military research projects in order to provide the scientific basis for future military applications.

If one looks at the military research infrastructure as a pyramid, then at the highest level would be the national laboratories and defence project agencies, which fund and coordinate military research projects and set the priorities for procurement. They are usually agencies within the national ministries of defence. The US has not just one but several national laboratories managed by the DoD, the Department of Energy (nuclear weapons) and NASA, to name the most important. The central DoD research organization is DARPA[1] (formerly ARPA), in the UK it is the DSTL,[2] in France the DGA,[3] in Germany it is the BWB,[4] in Sweden the FOI[5] and in Canada it is DRDC.[6] These national military research organizations conduct basic research, as well as applied research, which can later be developed into military technology for new weapons systems. There are also some international military research organizations, most importantly NATO's Research and Technology Organization (RTO), which is based in Brussels and dates back to the AGARD[7] created in 1952. The RTO was formed in 1996, combining AGARD and NATO's Defence Research Group (DRG), to improve the international cooperation in military R&D and provided advice and assistance to its member states for aerospace research and development.[8] On a lower level there are the laboratories of the different military services, which formulate the requirements for new weapons systems and which always have worked very closely together with the defence industries on the design of the next generation of weapons systems. For example, the US Army maintains several research labs, so does the US Navy, Marine Corps and US Air Force.

To a certain extent this is mirrored by the armed forces of other NATO countries. At the lowest level of the Cold War military research pyramid there are the universities

1 **D**efense **A**dvanced **R**esearch **P**rojects **A**gency, see www.darpa.gov.

2 **D**efence **S**cience and **T**echnology **L**aboratory, formerly DERA, see www.dstl.gov.uk.

3 Délégation Générale pour l'Armament, see www.defense.gouv.fr/sites/dga.

4 Bundesamt für Wehrtechnik und Beschaffung (Government Agency for Defence Technology and Procurement), see www.bwb.org.

5 Swedish Defense Research Agency, see www.foi.se.

6 **D**efence **R**esearch and **D**evelopment **C**anada, see www.drdc-rddc.dnd.ca/home_e.asp.

7 Advisory Group for Aerospace Research and Development.

8 Compare NATO RTO website www.rta.nato.int/Main.asp?topic=RTOHistory/AGARD.htm.

and other independent research institutes and organizations, which mainly did basic research without any concrete applications in mind. With respect to the US it is clear that many of the most prestigious universities have received a lot of funding from the Pentagon, among them Berkeley, MIT and CALTech. MIT even set up its own defence research wing in 1951, the Lincoln Laboratory, which has worked on many important military research projects like the development of satellite communications for national defence in the 1960s. In addition MIT and other universities served as a talent pool for the then expanding defence and aerospace industries. Also in other countries universities attracted military funding, but certainly to a lesser degree (both absolute and in proportion) than was the case in the US. For example, in Germany the amount of military research carried out by universities was negligible and still is. Eventually, the end of the Cold War marked a significant change in the size and organization of military research, which had many reasons, not only the general cuts in military spending.

In the US there is still a lot of military-related research going on, but the way it is organized has changed significantly. It is by no means as tightly government controlled as it used to be during the Cold War. Information on military research projects is now surprisingly accessible, at least as the official defence projects are concerned, while 'black' projects still leave room for a lot of speculation.[9] After the end of the Cold War many of the smaller labs were closed down and the responsibility for military research has increasingly shifted to industry, in particular to the prime contractors. The national defence labs have also cut back personnel and now have problems attracting qualified researchers (Repsher Emery 2002). They were compelled to hire contractors and services from the private sector in order to be able to fulfil their function. In the case of the Sandia Labs, they are now completely maintained and operated by Lockheed Martin, which uses the facility to conduct nuclear weapons research for the Department of Energy (Caldicott 2002, 37). The Idaho National Lab has pursued a similar approach and has partnered with a private company, Battelle, to conduct nuclear energy research.[10] Ann Markusen notes that 'Outsourcing of RDT&E, in particular, has increased dramatically and Navy technical centers outsourced 50% of RDT&E by 1996, up from 30% in 1970.' (Markusen 2001)

DARPA and also other national and services labs are now hardly able to do more than financially manage military research projects, which are contracted out to the industry, or to *Federally Funded Research and Development Centers* (FFRDC). FFRDCs are usually non-profit organizations like RAND, MITRE,[11] MIT's Lincoln Laboratory, or the Oak Ridge Laboratory and are now in many cases run by defence contractors or universities (Bischak 1999, 61). They are often in effect running the research projects managed by DARPA and also have close connections to the industry. The universities also collaborate today much more with industry than

9 David Hambling claims that the black budget for defence projects in the US was in the region $23 billion in 2004. Enough for Ufos, Vortex Guns and Mind Control Weapons, or whatever other sinister projects the US military wants to hide (Hambling 2005, 180).

10 Compare Idaho National Labs website, www.inl.gov/about/index.shtml.

11 MITRE operates three FFRDC for the US DoD.

with national governments, which is reflected in the funding they receive from the industry compared to overall government funding.

The prime contractors usually have their own research facilities and manage big defence projects in close cooperation with the military. Already legendary is Lockheed's *Skunk Works* near Groom Lake, Nevada, where revolutionary aircraft like the SR-71 and the stealth fighter were developed. The *Skunk Works* is a design bureau, which gives the engineers working there in small groups, a lot of freedom in their approach, so that they can operate independently, while managerial oversight is kept at a minimum, as the investment required is modest (Rich and Janos, 346). Engineers can play around with various designs and those that look promising in tests can be developed further. Boeing has followed Lockheed's example and has set up its own research wing named *Phantom Works*, which is developing new ideas and enable Boeing to make best use of its own internal and external resources. Only the best engineers and managers of Boeing work there and have access to all the technology (within Boeing and also from universities) and the venture capital necessary 'to turn revolutionary concepts into reality'.[12]

Very close to the defence industry and also peripherally involved in defence research are the political think tanks, which among other things evaluate the impact of new technologies or lobby for particular research projects like the necessity of a ballistic missile defence. Best known are probably the *Heritage Foundation* or the 'neocon' think tanks *National Institute for Public Policy* and the *Project for the New American Century*. These seem to have significant influence on the foreign and military policy of the present Bush administration (Hartung 2003, 91-115). Sometimes the connections between think tanks and defence industry are plainly obvious. So in the case of Northrop Grumman, which has now its own corporate think tank, the *Analysis Center*,[13] that wants to advise defence policy on a variety of issues like transformation, ballistic missile defence and other projects and technologies in which Northrop Grumman obviously has a vested interest. While the institution of the private 'think tank' is a prominent feature in the defence establishments in the English-speaking world, they are much rarer and much less influential elsewhere.

Also in Europe there were some significant recent changes in respect to military research. Perhaps most importantly, Europe has realized the continued need for high quality military research and also the need for a higher degree of technology cooperation in face of the budget cuts of the 1990s and rapidly increasing R&D costs for high-tech projects. The gap in military research investments between the US and Europe is quite huge, as the US accounts for half of the world's R&D spending (Paarlberg 2004, 130) and spends more on military research alone than Germany overall on its armed forces. Figure 3.1 compares US and European defence speaking, which indicates that the Europeans spend significantly less overall and far less on R&D. This development is from a European perspective quite worrying. The Europeans fear that their forces might not be interoperable in the future unless they purchase US systems (Pattie 1999, 57). Therefore, Europe is trying to make

12 Compare Boeing website, www.boeing.com/phantom/flash.html.

13 See Northrop Grumman website, www.analysiscenter.northropgrumman.com/index.html.

better use of its resources and integrate national defence industries better into a collaborative European effort.

Figure 3.1 Defence Expenditures Comparison US-Europe

Comparison Defence Expenditures US – Europe (2004)	**Defence Spend Allocation (In bn USD)**

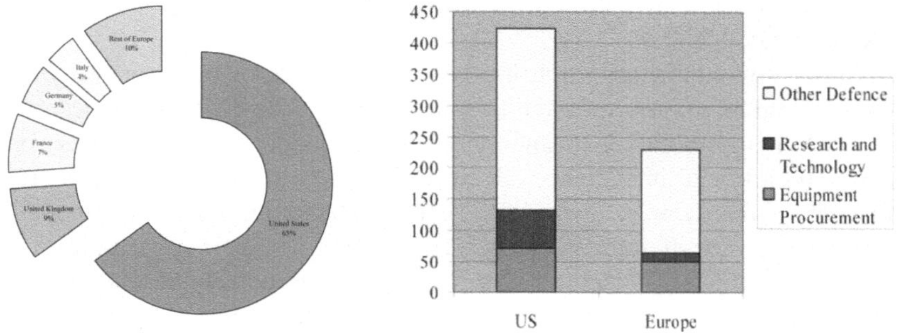

Source: BAE SYSTEMS, *Annual Report 2004*, p. 8.

In 2004 the European Defence Agency (EDA) was created in Brussels, which has the function to improve the European defence capabilities by improving and facilitating the cooperation of the national defence industries. The aim is to create a continental defence market, which can provide Europe with the military capabilities it needs.[14] EDA is intended to ensure that wasteful duplication of effort in European military R&D is stopped and that European forces remain interoperable. However, EDA is not a supranational body and can not dictate national procurement policy. At the moment its operational budget is just sufficient to fund feasibility studies (European Union 2005, 30). EDA is certainly a first step towards developing a combined European military capability independent of the US, but considering the tiny budget (€20 million) and few personnel (77), it remains questionable how much strategic direction it can achieve.

In Britain even part of the defence projects management, testing and evaluation has been outsourced to a private company (Qinetiq) (Briody 2003) when DERA was split into Qinetiq and DSTL in 2001. DERA has been privatized in order to improve knowledge management and transfer between government research labs and the industry. Qinetiq is now responsible for all military tests (Merindol 2005, 164). The majority of shares are owned by the British MoD, but about 34 per cent were bought by the American Carlyle defence investment group in 2003. Qinetiq itself has also strengthened its American ties and increased the spectrum of its business by acquiring two American technology companies, Westar and Foster-Miller. Apart from Qinetiq's testing and evaluation work for the MoD, it additionally does applied

14 Compare the European Defence Agency website, http://ue.eu.int/cms3_fo/showPage.asp?id=277&lang=en&mode=g.

research for developing military technology like radar-absorbing coating. Qinetiq also develops software for the defence and security industries. Its close affiliation with the industry and its American business raises serious questions about the impartiality of Qinetiq in evaluating defence projects, which might clash with its business interests. However, the MoD retains an independent RDT&E capability in form of the DSTL. It still operates several labs at more than a dozen locations in the UK, but its workforce is a lot smaller than under DERA (3,000 DSTL compared to 15,000 DERA) and it has already become a serious concern that all the competent people had left, while only the 'second and third rate' scientists stayed (James, Cox and Rigby 2005, 158).

Additionally, civilian research has become much more important, as civilian technology can be converted for military applications. Universities still receive a lot of money from the military to do important basic research, particularly in electronics, robotics and artificial intelligence, all of them also quite relevant for commercial applications. For example, in April 2004 the Pentagon held a public competition for the development of autonomous vehicles and promised the winning team prize money of $1 million. The competition was repeated in September 2005 (Hallinan 2004). On the other hand, there are lots of technologies, which remain military-specific and which have no civilian application or equivalent like stealth technology or tank armour, and which need to be developed by military research labs.

Trends in Military Research

There is little doubt that military research remains important because not every technology the military needs to stay technologically ahead can be derived from dual-use technologies. The continuing importance of military research can be seen in the Pentagon's eleven per cent increase in military R&D spending between 2002 and 2005 (UK MoD 2005, 25). There are also real opportunities to utilize commercial markets and technologies either by partnering with industries, or by military spin-off of commercial products, which allows savings on the military research side. For example, Los Alamos and Livermore had to employ lots of programmers for developing operating systems required for the supercomputers they used to simulate nuclear explosions. As supercomputers are now commercially available, there are now also commercially available operating systems for these computers, thus making it unnecessary for the research labs to develop their own (Beason 1997, 98). Another example is the *Affordable Weapons System* presently developed by Titan, which is completely based on commercial-off-the-shelf technologies and which has the features of a cruise missile, but at a tenth of the costs.[15]

However, there are also problems. The national defence laboratories in the US and Europe have really scaled down in terms of funding and personnel with the result that they do much less basic research. This is problematic as military research is now largely in the hands of the prime contractors, which have no real incentives for doing basic research. As a result the focus of military research is now on applying

15 See Titan Affordable Weapons System Public Relations material, available at www.titan.com/products-services/load_pdf.html?filename=337__1124001001.pdf.

already known science in order to quickly develop military technology, which is then rushed to the battlefield before it has been sufficiently well developed and tested. There are lots of cases where this happened. For example, JSTARS, the Patriot missile and Globalhawk have all been deployed before they had officially entered service and they had to be modified in the field to correct errors. This negligence of military research could widen the technological gap between the government and the commercial industries and might have negative long-term effects. These are some fundamental problems, to which there are at the moment no real answers.

The high number of potentially important technologies, their growing complexity and the limited national research resources will also make it increasingly difficult for military research labs to retain a basic competence in even the most relevant fields. Usually it also cannot be judged which technologies are worth investment and whether it is possible to reap any benefits from particular research projects. Previous misjudgements concerning the potentials of technology pitched by weapons manufacturers have sometimes turned the Pentagon into a laughing stock. A recent case brilliantly described by Sharon Weinberger is that of the 'Isomer' weapons. The Pentagon spent millions of dollars on researching a nuclear type of weapon, which could potentially equip a single soldier with the firepower to destroy a small town. However, it turned out that the concept of these weapons violated some basic laws of physics (Weinberger 2006).

Considering the costs involved, the risks are great, and without retaining sufficient scientific competence the risk for governments funding complex weapons projects will become even greater. The most illustrative examples are missile defence systems, where even experts are not sure whether it is technologically feasible at all. A reason for the shift in competence from the government research labs to the defence industry is that the growing complexity of weapons systems increases the role of the prime contractor in the innovation process. The prime contractor has the responsibility of managing all the subcontractors and often also for systems integration, which is an important activity in its own right.

Systems Integration

An important aspect of the R&D of new weapons is increasingly systems integration, as components and weapons systems need to function as an integrated system, thus requiring compatible technical standards. It can even be said that 'the key and most difficult aspect of defence work is likely to be the integration of complex systems, not the production of individual components' (Haglund and MacFarlane1999, 14). With the growing importance of commercial components for the new defence industry, the skills for integrating them into complete weapons systems are more crucial than the manufacturing side of it. Furthermore, it can be argued that system integration skills are at the core of the RMA and are the key for maintaining a technological advantage. The US DoD believes that 'U.S. military advantage will derive less from advanced component and subsystem technology developed by the U.S. defense sector than from the military functionality generated by superior, though not necessarily U.S. based, defense sector systems integration skills.' (US DoD 1999, 8)

With respect to the RMA, systems integration creates the basis for C4I systems or network-centric warfare. C4I systems are based on the idea of linking various command-, control-, communications-, computer- and intelligence systems into one system that manages all. Systems integration services and C4I systems have become a large market segment in the defence industry with further growth to be expected (Battilega e.a. 2000, 3). Jane's *C4I Systems 2004-2005* lists no fewer than 997 different command, communications, intelligence, and computing systems presently in use in by modern armed forces all over the world. This includes everything from the AWACS airborne command post to the small desktop application. The sheer number of different systems used within the armed forces and by the different countries makes interoperability a high aim. In the future it is hoped that the forces can depend upon just one system-of-systems, which would provide full interoperability. At the moment the various systems still need to be shuffled together in an improvized fashion. Not surprisingly, this is momentarily one of the most important fields where technical contractor support is needed and sometimes is crucial.

Types and Definitions

The key for the transformation to fully digitized armed forces is systems integration to ensure interoperability of the various forces and platforms. All the different high-tech systems need to work in concert, which means that the networked platforms can share information and interact with each other in real time. This requires compatibility of the systems and sufficiently good communications between them. Systems integration has therefore to start at the level of overall procurement planning in order to ensure that the many different systems used by the military and which are manufactured by many different companies are compatible to each other. It is therefore a separate task to designing and manufacturing weapons and platforms and is often performed by separate and specialized organizations.

The overall defence industry can be described as a pyramid with the systems integrator at the top. Below are the prime contractors, which are in charge of the overall management of defence projects, and further below are several layers of subcontractors (First Tier, Second Tier, Materials and so on). The main job of the systems integrator is to introduce and maintain technical standards, which are used by all systems and by all contractors and subcontractors (Gholz 2002, 5). On the highest level system integration would mean to define the overall network itself and then shape the different nodes of the network to fit these requirements, which can be called 'system-of-systems integration' or 'architecture systems integration'. On a lower level are weapon system integration and platform integration. Both of these can be equally complex tasks depending on the systems concerned. Weapon system integration is about integrating the different components from different suppliers into a single product, while platform integration concerns the integration of very different types of equipment into a weapons platform. Lower level systems integration is generally performed by the prime contractors and is less relevant for this thesis. Much more important is 'system-of-systems integration', as these system integrators play a crucial role in the process of transformation. They define the network, its components, the interaction of these components and their functions.

This is a very complex task, as it 'requires a detailed understanding of military goals and operations, and a sufficient trust to bridge military, economic, and political interests' (Gholz 2002, 7).

In the US there are several organizations and companies, which do high level systems integration. There are the military system commands SPAWAR, NAVSEA, NAVAIR, the National Laboratories and the FFRDCs like MITRE. Additionally, there are private companies like SAIC, Titan and CACI, which specialize in IT and systems integration. Finally, there are the big prime contractors like Lockheed Martin, which also try to take over more and more responsibilities in regard to systems integration. In Europe system integration is done on a national level by prime contractors, which causes problems in terms of interoperability of the various C3 systems used by the different European armed forces. This problem became obvious during the Kosovo air campaign in 1999. The coalition command lacked secure communications and the coalition partners had very different capabilities, which meant that the coordination of coalition attacks was frustratingly bad and that there was an increased risk of friendly fire. The new European Defence Agency is now tasked with integrating European defence capabilities. It has established a *C3 Flagship Programme*, which will take time considering the limited European defence budgets and the different national priorities (Ben-Ari 2005, 2-4). Recently there has been an international industry initiative to establish standards for network-centric systems. The Network-centric Operations Industry Consortium (NCOIC) wants to develop a 'globally interoperable architectural framework', so that platforms and other systems can be easily plugged into existing networks. NCOIC works together with NATO and the EDA.

In respect to weapons system integration the current trend is to hand over the task of systems integration to the prime contractors, which would become the Lead Systems Integrators (LSI). The primes would simply buy all the components for a new weapon system and use their expertise to put them together, which would reduce their investment and risk. This allows primes to take on contracts for items for which they have no own manufacturing capability. Lockheed Martin, for example, which has no shipyard, could bid for the US Navy's new Littoral Combat Ship. Sometimes it is even preferred to choose companies as systems integrators, which have no own production capability. This is the case for the UK's *Future Rapid Effect System* programme for which Atkins was awarded a contract 'to lead the Initial Assessment Phase' (www.atkinsglobal.com 2004).

Systems integration has become a market niche for US and European defence companies. Second-tier weapons suppliers are often not capable of it, as the entry costs for developing the capability are too high. As a result experienced systems integrators offer their expertise on the global market (Battilega e.a. 2000, 3). The Pentagon is worried that less developed countries could substantially enhance their military capabilities by hybridizing systems.[16] This means combining components from various countries of origins into much more effective systems (US DoD 1999,

16 E.g. India purchased T-90 tanks from Russia and then upgraded them with French Thermal Imaging systems to give the tanks a better night fighting capability (Dunnigan 2006).

23). It could also allow less advanced countries to quickly develop military high-tech systems by hiring commercial systems integrators for putting together new weapons systems as spin-offs of commercial off-the-shelf technology. The possibility was already demonstrated by Titan's *Affordable Weapons System* and the *Sea Fighter* shallow water ship, both of them put together with commercial products and technology for a bargain price (Associated Press 2005). There are some companies doing less technically sophisticated systems integration, which is more about military consulting for procurement decisions to make sure that various pieces of equipment fit together. Consulting companies like Booz Allen Hamilton offer services related to systems integration, not so much from the engineering side, but from the side of acquisition advice and advice for better integrating capabilities.[17]

Battle-management Systems

Battle-management systems provide military commanders with situational awareness and command and control capabilities and are installed in most stationary and mobile command posts. Sometimes battle-management systems are also referred to as Command and Control (C2) Systems or as Common Operating Picture/Common Operating Environment (COP/ COE), which allow multiple platforms in an area of operations to share, for example, the same radar picture. They usually consist of an operations centre stuffed with dozens of small computers, large computer terminals, and huge screens with maps and other information indicating the location, strength and status of own and enemy forces. Attached to them are usually sensors (radar, sonar, imagery etc) and communications systems (satellite, land line, wireless). They are at the heart of any modern military command centre and no modern armed forces can do without them.

There is now a wide range of stand-alone, joint-, land-, maritime- and airborne C2 systems in use by armed forces since the early 1980s and they tend to be very complex. Contractors are needed for the installation and upgrading, the training of military personnel, and for maintenance and trouble-shooting. The following section will describe some of the most important battle-management systems presently in use.

The very archetype of battle-management systems is the NORAD command and control centre in Space Mountain, Colorado, from where the US strategic forces would have been controlled during or after a Soviet nuclear attack. The NORAD battle-management system is equipped with artificial intelligence and was also used for planning and wargaming nuclear wars in order to develop a defence system that would have been so survivable that it could have fought World War IV after a six-month all out World War III (Gray 1997, 64). NORAD exists since 1963 and will remain important for the future thanks to plans for substantially upgrading it. Lockheed Martin has recently won a $1.5 billion contract for an upgrade of NORAD over 15 years, which will integrate 40 different C2 systems into one global battlefield

17 For example, Booz Allen Hamilton helped the US Air Force to integrate existing and future weapons to make operations more effective (Booz Allen Hamilton 2004, 20).

picture for air-, land-, sea- and space operations, including space warfare (Wolf 1999).

The US Theater Battle Management Core Systems

Following the experience of Desert Storm (and Vietnam) the Pentagon realized the need to bring all airpower under one unified command and information system. In Desert Storm there was not much interoperability of the C2 systems used by the service components. This proved to be a major obstacle to the planning and coordination of air operations. In Desert Storm the planning of sorties usually took several days, which allows really not much flexibility. This problem was eventually addressed with the development of a joint battle-management system, the Lockheed Martin *Theater Battle Management Core Systems* (TBMCS) introduced in 2000. TBMCS can reduce the time from sortie planning to execution to less than an hour.

The TBMCS integrates a whole range of different computer systems and functions in order to create a seamless C4I environment, namely the Contingency Theater Automated Processing System (CTAPS), the Wing Command and Control System (WCCS) and the AMC Command and Control Information Processing System (C2IPS)(Trivett 1996). The integrated functions include, according to Jane's Information Group, 'intelligence processing; air campaign planning, execution and monitoring; aircraft scheduling; unit-level maintenance operations; unit- and force-level logistics planning; and weather monitoring and analysis' (Ebutt 2004, 116). Additionally, the system has also several modelling and simulation tools for threat analysis and prediction, as well as automated targeting and sortie planning, which automatically selects targets and then matches aircraft and munitions to the targets. TBMCS is presently used by the US military and was recently introduced into service with the Australian Air Force. The battle-management system was used for carrying out the air war in the 2003 Iraq War. According to Lockheed Martin, TBMCS played a key role in 'planning, managing and executing the war'. Lockheed Martin claims that it helped to coordinate up to 41,000 sorties in three weeks and that it provided an accurate overview of enemy and coalition forces (Lockheed Martin 2004a, 4). Obviously Lockheed Martin contractors were needed to set up the computer network for TBMCS in the first place because it is a new system and the US military had no experience with it. David Isenberg points out that 'The services relied on civilian contractors to run the computer systems that generated the tactical air picture for the Combined Air Operations Center for the war in Iraq.' (Isenberg 2004) He also notes that 60 contractors followed the 4th Infantry Division into action to support other computerized C2 systems (Isenberg 2004).

Phalanx and Aegis

One of the most sophisticated battle-management systems is the Aegis Weapons System for fleet air defence, which is installed on American Arleigh Burke class destroyers and Ticonderoga class cruisers. Aegis was developed by Lockheed Martin, which also provides the technical services necessary for the maintenance and operation of Aegis. Raytheon and BAE SYSTEMS are also providing technical support services for Aegis. The system has been exported to Spain, Korea, Japan, Norway and Australia, and is installed on more than 71 warships. Aegis can track up to 100 aerial targets in excess of 100 miles, identify them and, if necessary, engage them. The Pentagon has recently awarded Raytheon a contract to upgrade Aegis for Theater Ballistic Missile Defense. A similar, but less sophisticated system is the Phalanx weapons system, which is installed on American and British warships and which consists of a radar tracking system and two 20mm guns directed by a fire control computer. Phalanx can automatically identify and engage aerial targets. The system is now installed on nearly every class of American warships. Overall, Phalanx and Aegis seem to be now a success story, but these two systems are also very good examples for the potential risks and problems such highly sophisticated systems can cause under operational conditions. In fact the failure of Phalanx has led to the near loss of the American Perry class frigate USS Stark in 1987 and the failure of Aegis to the accidental downing of a civilian aircraft killing 290 people in 1988. What happened?

According to Chris Hables Gray, Phalanx failed because of known technical problems, which caused the system to only run 6-12 hours before breaking down. The contractor General Dynamics had used special software to override a Navy testing program and the crew of the USS Stark was obviously not aware of the defect (C.H. Gray 1997, 65). When an Iraqi aircraft accidentally attacked the ship with an Exocet missile, Phalanx was already down and could not intercept it. As a result 37 sailors lost their lives. Having learned that lesson, Commander Rogers of the Ticonderoga class cruiser USS Vincennes decided not to take any risk when the ship's battle-management system indicated an Iranian F-14 descending to an attack position. He ordered to engage the target with two sea-to-air missiles and it was destroyed. The aircraft turned out to be an Iranian civilian airliner on a scheduled flight, which was perfectly on course and ascending, not descending. Aegis was during that time not in an automatic mode and the crucial mistake was made by an operator, who mistook the airliner for a nearby friendly F-14 and designated the aircraft incorrectly. However, there are two important points to be made: first, although Aegis was still a man-machine system, which allowed human interference at all times, the system automatically inserts (manually designated) targets in an engagement queue and engages them without being directly controlled by human operators (C.H. Gray 1997, 66); secondly, the interesting thing about the incident is not that Aegis itself failed, but that the system was simply too complex for the human operators (Swartz 2001). For this reason it might make sense to let the system operate completely on its own instead of having humans interfering with systems they do not completely understand (Adams 2001, 66). For the time being, humans are still in the loop, but at the same time it is clear that interference in a complex system like Aegis requires a high level of technical skills and understanding of the workings of the system, which is hardly available in the US Navy or other navies. This is illustrated by the fact that during the recent Iraq War contractors had to help operating the Aegis systems on US warships (Singer 2005, 119-132).

Ballistic Missile Defence

Compared to Aegis, BMD systems are even much more complex because of the extremely difficult task for which they are developed. Missile defence systems consist (like Phalanx and Aegis) of a battle-management system that identifies and tracks targets and a weapons system that can independently engage these targets. Because of the speed of the targets (ICBMs fly at speeds of 15,000 mph) and the need to engage them quickly before they can become a threat, any ballistic missile defence system will have to be autonomous and therefore extremely complex, as this requires artificial intelligence. The difficulty of developing a system, which can detect targets and make an independent 'intelligent decision' on what target to engage and which is also precise enough to effectively destroy very fast moving targets, should not be underestimated. No wonder that the present success rate of BMD tests is not very high. On the other hand, it appears that missile defence systems will and can be built. In this case it is already certain that the companies that develop these missile defence systems will be needed to conduct, or supervise the maintenance and operation of these systems, as malfunction could mean that wrong targets are engaged, or real targets are not intercepted.

MEADS is an air defence system capable of intercepting medium range ballistic missiles, which is being developed by Lockheed Martin and its European partners. MEADS will replace the Patriot, Hawk and Nike Hercules systems presently in use by the US, Germany and Italy. MEADS can be used as a fully integrated air defence system for homeland, as well as theatre missile defence. It can not only control its own PAC-3 based missiles, but potentially also other air defence systems/sensors and shooters. This will provide more firepower with much less manpower. It can be easily transported and deployed and once activated it can operate largely autonomously. If there are no major technical or political obstacles, MEADS could be fielded by 2012. Considering the fact that contractors were needed to maintain and operate Patriot systems in the recent Iraq War, the likelihood that Lockheed Martin contractors are needed to perform similar services on the much more complex MEADS is pretty high.

The Airborne Laser System (ABL) currently developed by Boeing will be able 'to detect, track and destroy missiles in the boost stage', and will be based on a Boeing 747-400 aircraft. The weapons system consists of a chemical oxygen-iodine laser (COIL), which fires a high energy beam at the target and is controlled by a battle management and integrated beam control/fire control system (Boeing 2004a). The system has to calculate the speed and trajectory of the attacking missile and would point the laser at a predicted impact point, while correcting tilt and distortion of the beam caused by atmospheric conditions. The aircraft would fly at an altitude of 40,000 feet and the airborne laser would be capable of autonomous operation. Should the project proceed according to plan, then there should be seven ABL systems ready by 2008 (see www.fas.org). The downside of the ABL are its extremely high development and operating costs. For the system to work ABL aircraft would need to be airborne all the time, would need to be protected by escort aircraft, would depend on tanker aircraft for extended range and would have extensive maintenance and support requirements. GAO therefore estimates the operating costs at $24,000 to $92,000 per hour (US GAO 2004). Keeping a fleet of ABL aircraft operational would consume several billions per year.

Network-centric Solutions

Many modern armed forces are planning to introduce network-centric systems that would allow interoperability and interconnectivity of all military systems within the next decade or so. Even less advanced forces such as the Indian Army are discussing the potential benefits of network-centric solutions. It is quite likely that once the most modern armed forces have been successfully and completely digitized, a whole lot of others will follow. At the moment the problems of introducing big computer systems into the armed forces, which would manage and control nearly everything, are substantial. It is very likely that a successful overall digitization of the armed forces will take a long time, maybe even to the middle of this century.

The US has operated several domestic and global military computer networks for many years. Chris Hables Gray claims that they were in the early 1990s 'organized into at least nine metasystems: (1) The Worldwide Military Command and Control System (WWMCCS); (2) the Modern Age Planning Program (MAPP); (3) the Joint Deployment System; (4) the Stock Control and Distribution Program; (5) the Tactical Army Combat Service Support Computer Systems (TACCS); (6) the Military Airlift Command Information Processing System (MACIPS); (7) the Airborne Battlefield Command Control Center (ABCCCIII); (8) the Joint Surveillance Target Attack Radar System (JStars); and (9) the Operation Desert Storm Network (ODS NET)' (C.H. Gray 1997, 38). The US has also ambitious plans in respect to replacing these networks in the future in order to create fully digitized and networked armed forces. Presently Lockheed Martin is working for the Pentagon on a Transformational Communications System, which will provide the US military with a flexible, highly mobile global wireless communications system (www.lockheedmartin.com 2004). Eventually the US Army will field a fully digitized and integrated force of at least 14 *Future Combat Systems* (FCS), which are still at least a decade away.

Transformation Support

The first step in this direction was taken with the Task Force XXI advanced warfighting experiment at Fort Hood in 1997 where the digital battlefield was put to a test. This resulted in the decision to go ahead with Stryker Brigade, which is a completely digitized unit and which has already been fielded in Iraq.[18] Stryker Brigade has proven that it can be very effective in the difficult urban environment of Iraq and the project is considered to be a success. The big question is, whether normal units can be equipped with the technology and whether they will be able to handle it. Christopher Toomey points out that 'When it comes to digitization,

18 Stryker Brigade consists of 19 ton lightly armed wheeled armoured vehicles, which come in 10 different configurations. Special about Stryker Brigade is its three times bigger reconnaissance battalion equipped with UAVs and realtime datalinks. The Strykers can operate 48 hours without refuelling and can drive 60 mph. (See Freedberg 2003).

civilian contractors fielding the systems or their numerous modifications do most of the training. Digital expertise, the knowledge to effectively use the systems in combat, rests with highly talented contractors, not with uniformed soldiers. There is no influx of trained, digitally aware soldiers, and this will not happen in the foreseeable future.' (Toomey 2003, 43) This does of course apply to all other armed forces, which are about to introduce network-centric systems. Concrete plans for force digitization exist in Britain, Germany, France, Australia, the EU and NATO.

Britain is pursuing a 'Network-enabled Capability' and is currently negotiating with several companies to build a Defence Information Infrastructure (DII) (C4ISR Journal 2005). The system will be built by a consortium named Atlas and the prime contractor will be the American computer services company EDS. EDS will work together with the Japanese Fujitsu, the American General Dynamics, the German/French EADS and the British LogicaCMG. DII is planned to be introduced in three stages. The first stage is projected to cost £2.4 billion and will only cover about 70,000 desktop computers in offices, bases and Royal Navy warships. After successful completion, the second and third stages will include more personnel and systems until a single computer system can connect 340,000 personnel in 2,000 locations in 24 countries. Contractors will play a crucial role not only for developing and introducing this large computer system, but also for maintaining and operating it. Atlas will provide the DII support for a ten year period with an option for further five years.

'Bowman' will be the computerized wireless communication system, which will connect British army forces in the field through secure communications (see www.army.mod.uk/bowman 2006). It will be equipped with GPS and a data exchange system for generating a Common Operating Picture. It is planned that 18,000 army vehicles will be equipped with Bowman, which will make it necessary to train all 60,000 Royal Army soldiers on the system. The prime contractor is CDC Systems UK Ltd. (a General Dynamics subsidiary), which developed Bowman together with ITT Defence, BAE SYSTEMS, EADS and many others for a price of €2.6 billion. The system was introduced into the Royal Army in 2004 and the procurement process is scheduled to be finished by 2007.

The German Army is trying to catch up, but is still far behind the American and British forces in terms of implementation for political, rather than technological reasons. The technology for the digitization of the German forces would be available, but it will take a very long time to implement it, as the 1990s were for the Bundeswehr in terms of IT a lost decade. Most army communications still go over old fashioned telephone lines and most planning is done with a marker on a paper map rather than with a digital system. The German forces utilize digital C2 systems, but do not have a cross-service wide C2 infrastructure. For tactical communications the Automated Corps Network developed by Siemens was introduced in 2000, but it has limitations in terms of bandwidth and interoperability (Adams e.a. 2004, 60). To address some of these problems, the German forces have recently awarded a CSC-led consortium with the name ISIC21 a $6.1 billion project to upgrade the IT infrastructure of the German forces over a ten year period. Other companies in this consortium include MobilCom and EADS (Defense Daily 2002).

First steps have already been taken for meeting the German armed forces most urgent operational needs. EADS is presently building the infantryman of the future (*Infanterist der Zukunft*) systems for the German Army, which includes a bullet-proof vest, NBC protection, night vision, digital navigation and communication and tactical speech and data communication. Germany has already fielded a test version of the system for operations in Kosovo 2002 and has ordered 196 systems for its Special Forces to be introduced between 2005 and 2007. Apart from that Germany takes part in developing and utilizing new EU and NATO C2 infrastructure, for example EU's new C2 facility at Northwood UK and in the re-organization of NATO's command structure, managed by the Allied Command for Transformation (ACT). Other Western armed forces, most notably Australia, which has also recently introduced the JP 2072 Battlespace Communications System (Land) for network-centric operations, are also working on plans for the introduction of digitized communications systems (Evans 2004b, 47). There is little doubt that developing, providing and running these network-centric systems will be a very profitable business segment of the defence and IT industries in the coming decades.

Military Information Technology

Information Technology has already tremendously transformed commercial business activities, for example, through fast transfer of information and money (investment) and new ways of marketing (e-commerce). Now IT will have a similar huge impact on military activities, as computerization is a key element in the transformation of the armed forces. To a certain extent this has already happened, as the armed forces now routinely use computers and computer networks on an everyday basis. At the moment there is still a lot of improvization because the integrated network-centric systems have not yet been fielded, or still need to be used in conjunction with older systems until the process of transformation is complete. As a result, off-the-shelf commercial systems are being utilized until the transformational systems are ready. The number of computers is increasing in the military world at a similar speed to that in the commercial world. However, the necessary computer skills for working with these complex computer systems are clearly not widespread enough in the military. As the armed forces in most cases use computer systems similar or identical to commercial products, it is quite natural that they have turned to private companies for help. With the growing importance of IT for the military a lot of private contractors are needed to set up, maintain and operate computers and communications networks, and it implies that IT contractors are often needed in theatre. The overall volume of IT services is huge: about $35 billion awarded to the private sector annually – and growing ten per cent to 15 per cent a year (Schwartz 2003). Defence contractor Lockheed Martin claims that 25 per cent of its sales derive from IT solutions and services (Lockheed Martin 2003, 3), but there are also lots of non-defence and smaller companies in the military IT market.

Computer Services

Not all soldiers are computer-literate and many of them require training in this field. Private companies and freelancers offer basic computer training to the military and do all the basic computer work like installing software, setting up LANs, or installing new hardware components. The significant involvement of private companies in such work only shows that there are no substantial or even any differences between commercial computers and those used by the military. According to John Parker, the Pentagon had 2.1 million computers at the end of the 1990s and 95 per cent of them utilized commercial software (Parker 2000, 87).

Military computers might have some specialized applications with no equivalent in the civilian world, or the computer themselves might be 'ruggedised,' so that they can be used in dusty, dirty or harsh environments, but essentially they are the same as commercial computers and any person, who has sufficient computer skills in the civilian sphere qualifies for the same kind of work in the military sphere as well. There is little spectacular about outsourced computer services, especially if these services are provided domestically. Sometimes computer services need to be provided in the dangerous environment of a war zone. For example, a San Diego computer company named Tapestry Solutions was hired by the Pentagon to install software in Kuwait, where two of its employees were ambushed and killed in January 2003 (Bredemeier 2003).

The rapid proliferation of small computers can be seen in the example of a small American computer company (GTSI), which claims to have sold ten thousands of ruggedised small computers to the military in 2002/03. This means that 'Every Army Ranger parachuting in [to the battle theatre] will likely have a Panasonic laptop attached to him.' (Rogers 2003, 5) In the future the US Army wants 'to go big' on mobile computing and has already allocated $5 billion for laptops over the next ten years (Onley 2005). There are already thousands of computers in Iraq, which need to be serviced and repaired. Numerous unidentified small private companies are involved in this work, but also some big computer companies like Dell, which holds a $500,000 contract to supply the US Army with computer services (Spearin 2003, 32). More than 500 IT contractors support the US occupation force in Iraq. The US Army has realized its dependence on IT contractors: 'Army officials who manage IT programs and efforts said they desperately need support from contractors.' (Tiboni 2004) A company that provides computer services to the US military in Iraq is Technical and Management Services Corp. (TAMSCO). The company mainly 'installs and maintains voice, video and data systems for the CPA. It provides those services using commercial satellite systems' (Tiboni 2004). Other companies like the Engineering Support Systems Inc. have already plenty of experience with this kind of work gained during US military operations in Haiti and the Balkans. Of course, there are also the big names supplying computer services in Iraq: General Dynamics, Lockheed Martin, Northrop Grumman and Raytheon.

For establishing the CENTCOM Tactical Operations centre at Camp Doha, Kuwait, 65 computer servers, which use the Secret Internet Protocol Router Network (SIPRNET), had to be set up and maintained by 50 military personnel. However, if they had trouble with the technology they could rely on Microsoft Online Help to which CENTCOM has subscribed to Premier Help, according to an article in the computer magazine *Wired* (Davis 2003). The troubled soldiers could then talk to a Microsoft live operator back in the US, who would give them technical advice, sometimes as simple as to change the batteries. According to the same article, SIPRNET enabled the US soldiers to use a Microsoft chat programme and to chat to others even in the middle of a battle. Microsoft Chat automatically assigns icons to users (if they are not chosen beforehand) so that 'a bunch of field artillery colonels will come online in the form of these big-breasted blondes' (Davis 2003).

There is not only Microsoft, which provides support for software, but it is now common practice in the software industry and as the military purchases commercial software, it can buy and utilize the support packages for it as well. The conclusion reached by CACI's CEO Jack London is that 'the role of many IT companies in supporting deployment has changed significantly' and CACI 'has been involved in setting up communication centers and network operations centers, as well as surveillance and sensor equipment for intelligence' (Wait 2003).

Network Security

Internet security is already a major business opportunity. In 2000 it was estimated that the worldwide Internet security market has reached a value of $5.1 billion and according to projections it would surpass $14 billion in 2005 (Denning 2003, 39). As the armed forces are becoming more and more dependent on digital communications and more weapons systems are connected to worldwide computer networks, network security is becoming a major issue. Dependency on technology inevitably creates vulnerabilities for the forces which use it. The military is concerned about the possibilities of Information Warfare, especially about computer network attacks against military networks. At the moment the US forces are investing a lot of money in network security, as they believe to be most vulnerable to hacking. Several successful hackings into Pentagon computers seem to justify this concern. Pentagon computers are attacked on a daily basis with thousands of attacks every year. Some of them are serious and hackers indeed managed to get access to information or to disrupt Pentagon computer systems (Dowd 2003).

A *Business Week* article claims that theoretically during the Iraq War any 'flyboys (and girls) could hop onto a special Air Force network from any PC equipped with a Web browser and special military encryption and authentication software. Once on this network, they could call for air strikes, direct reconnaissance planes, or plot the movements of the most powerful flying force on Earth – all from their laptop in a café (or, more likely, at a secured facility)' thanks to Lockheed Martin's *Theater Battle Management Core Systems* (TBMCS) (Salkever 2003). This possibility is too scary to let it happen. The Pentagon believes that the private sector has the best cyber weapons against these new threats and there are several big defence and computer companies specializing in computer network security in the government/military sector.

In the American market for network security there are many big names like CSC/DynCorp, L3 Communications, CACI, Titan and MANTECH, some of which also work for foreign armed forces. The principles and technology are quite similar to those used in the commercial sector: developing and setting up intrusion detection software such as firewalls, to have software against malicious programmes and use backups and repair programmes to recover damaged data. Because of its higher security needs, the military wants particularly sophisticated and customized software. One example is the '*data resilience in information warfare*' software developed by Northrop Grumman branch Logicon, which can detect network attacks and recover damaged data (Erwin 2000). CSC provides the US government and military with extensive computer services, among them computer security, network security and information security. The company has contracts with the US federal government worth $4.7 billion and many of them with the Pentagon ($2.9 billion) (CSC 2005, 16 and 26). Among them is the *Safeguard Program*, which aims to protect critical infrastructure of US federal agencies against cyber-terrorism.

Military computer networks are getting ever more complex, as the number of nodes and transmitted data grow at a rapid pace, making the military call for ever more bandwidth. Growing complexity and high data transmission rates do not only make the military networks slow, they also increase the risk for occasional disruptive computer crashes and successful computer network attacks. Despite tremendous efforts to ensure stable computer networks and data transmission capabilities, there is apparently no way to avoid a computer crash every now and then. Occasionally the problem is not solved by simply hitting the reboot button. More serious than technical errors can be deliberate attacks. All hackers know that the more complex a system, the more likely it is to find a backdoor into it. The American military (and to some extent other militaries as well) are about to create computer systems of a gigantic complexity and one could ask how long the defenders of these networks can keep up with the cyber-attackers.

Satellite Communications

Satellite communications have been used by some armed forces since the 1960s and until the 1990s it was the exclusive domain of the superpowers and less than a handful of other nations (Britain, France, China and Israel). Today many nations have satellites in space, or can take advantage of commercial communications satellites. Militaries have now become very dependent on satellite communications for managing ever growing communications loads necessary for their global operations. The US military has currently capacity of a data transmission by satellites of more than 1 Gigabit per second (GBPS), but it is expected that more than six times this capacity will be required by 2010 (MacLean 1999, 193). As satellites are quite expensive and capacities are not needed all the time, many modern armed forces are looking for innovative approaches for cooperation with the space and IT industries. Private companies are already involved in supporting military satellite communications in many ways. They develop and build the satellites, launch them, operate them from ground stations, and help setting up communications networks in theatre.

In the US the dominant players in the field of military satellites are since the late 1950s TRW and the Lockheed Missile and Space Corporation (LMSC), which built the first earth observation (Corona) satellites (see e.g. Taubman 2003). LMSC won in 1983 a contract to build a jam resistant communications satellite for the

US nuclear forces termed Milstar and in 1992 it also got the follow-on project Milstar II (launched in 1994). Milstar consists of four communications satellites in geosynchronous orbits and provides the US forces with secure voice, data, teletype and facsimile communications (Ebutt 2004, 235). In both projects TRW was the most important subcontractor. Milstar II has a much higher data rate capability than Milstar I, but to a very high price: project costs in 1994 to 1999 were in excess of $12 billion (FAS). Lockheed Martin also built and operates the ground stations to control these satellites.

Because of an unlimited appetite for bandwidth the Pentagon anticipates that its own satellite communications capacity will be insufficient and intends to lease commercial satellites whenever additional capacity is needed for contingency operations. During the 1991 Gulf War 25 per cent of US military communications travelled over the commercial Intelsat (Seebode Waldrop 2004, 165). Since 1997 the Pentagon has begun committing itself to the use of commercial satellite communications in the long-term in order to augment own capabilities (Rayerman 2003, 54). So despite the fact that the Pentagon has about 35 communications satellites in orbit worth $20 billion, it had to buy $400 million worth of satellite communications services from private providers in 2003 alone (Menke 2004).

Private companies also provide assistance to the militaries in setting up satellite communications and maintaining satellite terminals. Most of the 500 IT contractors presently in Iraq do work connected to satellite communications. Current military satellite capabilities are already insufficient for dealing with the enormous amounts of data that travels over satellites. Therefore, some countries have already leased commercial communications satellites for military operations and even the Pentagon is seriously considering making more use of commercial space assets, as future demand is uncertain and the operating costs for satellites are high.

Britain and France were originally the only nations with their own military satellite communications apart from the superpowers, but others have already followed (Spain-HISPASAT, Germany-SATCOMBw) and it is quite likely that more European nations will either develop their own military communications satellites, or become part of a future European military satellite communications network. Britain launched its first military communications satellite Skynet 1A in 1969 and has since then orbited several more advanced successors. Recently the British Government has decided to use the tool of a Private Finance Initiative for developing and operating a successor system to the present Skynet 4 satellites. EADS got a 15-year contract (to 2020) worth £2.5 billion for providing secure satellite communications to the MoD (EADS 2004a). Skynet 5A and 5B are scheduled to be launched in 2006 and 2007 (Ebutt 2004, 220). Paradigm Services, an EADS subsidiary, will also build, maintain and operate a new satellite control facility for the Skynet 5 satellites. According to Alistair McLean, the use of a PFI in the case of Skynet 5 is radically different from traditional procurement. He claims Skynet 5 'does represent an unprecedented move towards privatization of an almost front-line military capability' (McLean 1999, 196). As Paradigm/EADS will be the owner of the whole communications system, there is a risk of the company withdrawing the service prematurely for whatever reason, leaving the MoD without its global communications network. In peacetime this could be quite an inconvenience and it would take time to get a replacement working. In wartime it could be catastrophic.

EADS is also about to provide the German forces with its own satellite communications in order to support the global operations, which are now part of the new role of the Bundeswehr. SATCOMBw was launched in 2006 and is scheduled to be fully operational by 2008. In contrast to the British, the German military wants to have complete control over the satellite system, which is a significant change to the initial plan to only lease contractor-operated capabilities (EADS 2004b). Be that as it may, EADS and other contractors will play a key role in operating the military communications satellites, as the German armed forces simply have no experience with satellite systems and will certainly not be able to develop the necessary skills and expertise any time soon.

Canada has pursued a similar approach as Britain and the DND has awarded the SNC-Lavalin PAE (teamed with the Harris Corporation) with a C$200 million contract to supply communications and information services to deployed Canadian forces in 2002. This is part of the Canadian Forces Logistics Augmentation Program or CANCAP which has the function to support deployed Canadian forces with contractor-provided secure communications. According to Harris Corp., the company will 'supply specific packages such as communications hardware and satellite communications services' (Harris Corp 2003). This Canadian DND contract appears to be rather small compared to the national security business Harris Corp. has in the US. The company has recently been awarded a ten year contract worth up to $1 billion by the US National Reconnaissance Office (NRO) to provide maintenance and support services for the NRO's global communications network, which even includes day-to-day mission support (Harris Corp 2004).

Conclusion

The chapter focused on three closely related, but also distinct types of military services and activities, which have become increasingly privatized during the 1990s. Military research has acquired in the twentieth century a prominent role in the military sphere, as technological superiority could grant a decisive advantage in warfare. During the Cold War research spending amounted to almost half of the overall military procurement spending and this ratio has remained relatively constant. However, the organization and management of military research has changed significantly since the end of the Cold War. The national and military research labs now lack after years of downsizing the competence and manpower to independently carry out research. As a result, much of it has been outsourced to contractors and the private industry at large. Especially the role of prime contractors in military research seems to have become more extensive. In the case of Britain the main wing of military research has been privatized in the form of Qinetiq, a company which is partially foreign owned. In the future national governments will depend much more on private expertise for identifying defence relevant technologies. The growing range and complexity of new technologies combined with a growing dependency on outside expertise has already led to expensive misjudgements, which could become a more serious problem in the future.

Systems integration, or the ability to combine various components and systems into integrated systems or networks, has become a key capability for information age warfighting. It is the basis of network-centric warfare. The overall goal of Western armed forces is to integrate all military units and capabilities into a single system-of-systems. This will mean to insert wireless digital communications systems into any platform and military equipment down to the individual soldier. At the moment, various military C4I systems need to be shuffled together in an improvized fashion until the transformational systems are available. Because of the complexity of the task contractors are needed to make present military computer networks work. In the long run huge industrial consortia will develop, maintain and partially operate military information systems. As a consequence of the extensive private sector role in this area military systems integration is already available on the global market.

Finally, the chapter looked more generally at military IT services. While private contractors largely do the donkey work of setting up and maintaining tens of thousands of military computers around the world, they also work in more sensitive areas such as providing cyber protection against information warfare attacks. Particularly important is the military use of civilian information infrastructure such as telephone lines and satellite communications, which increases the vulnerability to disruptive attacks. Occasionally the military is even leasing satellite communications capabilities from commercial companies. In Britain it goes as far as the military leasing of privately owned military communications satellites. This already caused some concern whether such privately provided capabilities would be available in times they are needed most.

Chapter 4

Modelling, Simulation, and Wargaming

Modelling and Simulation (M&S) has been identified by the Pentagon as one of the 20 key technologies, which will ensure American technological superiority in the future (Schuppe 1991). Growing complexity of equipment, tighter defence budgets and a tremendous growth in computer capability have led to an increasing importance of M&S in the US and elsewhere. In fact, modelling and simulation has recently become the greatest growth area in the commercial, as well as the defence industries.[1]

What is it about? A computer model allows one to look at the features of a design or situation without the need of creating the actual machine. A model is a representation of an entity, which contains the relevant features. A simulation consists of series of models representing a system or organism, which contains the relevant behavioural features (Hambleton 2005, 251). In other words, it allows an examination of the whole process or development of the system or organism over time. Modelling and simulation are used in many contexts and for many purposes. For the military there are four main applications: 1) the development of weapons systems, including testing and evaluation; 2) military training; 3) combat analysis and wargaming; and 4) battle-management.

This chapter will outline the role private companies have assumed in respect to developing and testing new weapons systems, carrying out military training, planning and doctrine development. All these activities tend to be based on complex computer simulations, which is usually the reason why private companies get increasingly involved in these important aspects of military work in the first place. The following section will be about weapons design, testing and management activities.

Weapons Development, Testing and Management

The origins of computer modelling and simulation (and indeed of computers themselves) are their use for the development of 'the bomb' and the subsequent efforts to better understand the effects of nuclear weapons and to improve their features. The first problem the first American computer ENIAC had to solve was the theoretical problem of developing a hydrogen bomb (Fitzpatrick 1998, 10). The necessary calculations were simply too difficult to do manually. Computer simulations were later used to analyse the effects of nuclear weapons under various conditions and how this relates to shock and blast effects, radiation, EMP, and

1 According to the *National Defense Magazine* the market for visual simulations (computer graphics based simulations; usually used for training devices) grew almost 20 per cent from 1999 to 2000 to altogether $2.33 billion and is expected to grow to $5.3 billion by 2007 (Irwin (2001).

biological effects. The information was compiled in the US government publication *The Effects of Nuclear Weapons* in 1957 and updated several times (US DoD 1977). Since these early days of computation, when computers were mere instruments for developing the H-Bomb and simulating nuclear weapons effects, the applications for computer modelling and simulation have become much more numerous. It can help to design all kinds of weapons, not just nuclear ones. There are nowadays virtually no new weapons which are not developed with the help of computer modelling and simulation.

Weapons Design

Modelling and simulation tools are increasingly used for the development, the testing and the management of complex weapons systems like sensors, missiles, new platforms, and perhaps most importantly, of nuclear weapons. In pre-computer times it was a common practice to design a weapon on the drawing board, to assemble a prototype and then test it for real. If testing showed any flaws, the prototype would be modified and repeatedly tested until these flaws were overcome. Today weapons development and testing requires complex modelling and simulation because the weapons themselves are much more complex, which means that there are many more parameters, which need to be taken into account. When it comes to larger platforms like aircraft or submarines, it is also simply far too expensive to build a prototype only to find out that there are some major design flaws, which might only be noticed after production has started. By using simulation tools and computer models it is possible to test the whole system and all the components in advance and then optimize the functionality. Once the simulations show that the design of the system works, a prototype can be developed and used to correct some minor flaws. This makes the development of new weapons much cheaper and also faster because it is a lot easier and faster to change a few parameters of a computer simulation than to build and modify the real thing (Gansler 1998). Computer simulations also allow analysing processes, which happen either in a very short time frame (e.g. explosions or collisions, where it is necessary to see what happens within milliseconds) or over a very long time frame (e.g. the decay of radioactive material and effects of radiation – processes which take hundreds or thousands of years), which would otherwise simply not be possible.

The Pentagon (and also other governments) has introduced a new Simulation Based Acquisition (SBA) policy because of these advantages. The military contractors have to develop computer models and simulations in order to address not only design problems, but also the problem of projecting costs and duration of weapons programmes, issues of maintainability and life-cycle costs, and training aids.[2] For example, the design of the JSF could be optimized in terms of maintainability and life-cycle costs by using computer simulations. Delmia Corp. (a subsidiary of Dassault Systèmes) has developed simulations of maintenance work for Lockheed Martin's JSF, which has made sure that maintainers can reach any

2 See US DoD/PEO STRICOM website, www.peostri.army.mil/PRODUCTS/SBA/definition.jsp.

part and that components can be easily removed and replaced before a prototype was built (Thornton 2001, 204-209). Another consideration was to keep the overall maintenance costs for the JSF low and also the amount of necessary maintenance hours and the numbers of maintainers necessary. This is important as most costs of weapons systems relate not to the acquisition cost, but to life-cycle costs (which are at least twice as high) (Kutner 2001). Secondly, the number of qualified maintainers, both military and contractors, is dropping and without enough maintainers aircraft and other weapons spend more time on the ground or in the base and are not available for operations (Thornton 2001, 205). Apart from the great benefits the armed forces can get through modelling and simulation for weapons design, there are also problems and unaddressed issues. Ideally, the armed forces can work through SBA with the industry much more closely and optimize a weapons design before the prototype is built. The customer can look at the computer models and can make suggestions for improvements. Simulations can then determine how design changes affect the overall functionality.

This, however, would imply the sharing of data and simulation technology between the developer and the customer and would require a lot of trust. The companies naturally fear that they may lose their competitive advantage, if they have to share the technology of their design and simulation tools with the government or even with other companies (Kutner 2001). At the moment the Pentagon has no consistent policy in respect to SBA and it is expected that the problem will have to be solved once it becomes necessary not only to simulate single weapons systems, but a whole network of integrated weapons systems. This will be the case with the *Future Combat System* programme that connects 14 different platforms ranging from an unmanned howitzer to a man-portable UAV into a single network (Kutner 2001).

Taking this one step further is to develop a completely new generation of weapons on the computer, make them ready for production, but then put the blueprints on the shelf and start developing the next generation, which is the concept of 'skipped generations' (Defense & Aerospace Electronics 1993). If a new weapon is needed, it can be easily produced from the blueprints. Otherwise it is never produced and there are no costs of maintaining and operating unnecessary weapons, while weapons design can always include the newest technologies. In other words, governments could have 'virtual arsenals' of conventional, or more likely, non-conventional weapons. The obvious problem with this is that weapons systems are never sufficiently tested, only in computer simulations, which can have varying degrees of accuracy. If a new weapon has to be rushed into battle, there is no operational experience available, which turns the weapon into a liability rather than an asset. Under operational conditions a military commander will naturally prefer battle proven weapons and will be reluctant to grant any new weapon a more extensive role, which makes the idea of skipped generations rather questionable. What use is a weapon, which is not sufficiently tested? This leads directly to the next problem: the process of weapons testing and evaluation.

Weapons Testing and Evaluation

Weapons testing and evaluation (T&E) relates to activities which are intended to ensure the quality of equipment and reduce the risks for the armed forces. T&E is usually done by the armed forces- and national research labs, but because of reasons outlined in the previous chapter, the research labs now tend to contract out these activities to private companies.

In the US IT companies like SAIC, Titan, CACI, Anteon, and MITRE offer services relating to weapons testing and evaluation. In Britain the majority of test facilities are now operated by Qinetiq (UK MoD 2005, 10). In some cases the complete T&E of a new weapons system is contracted out and the contractor takes care of everything required to field-test it and will produce an evaluation report containing its findings. Few companies have the capability and expertise for a complete test. Usually a contractor only tests components of a weapons systems, or certain aspects of it. Some companies just provide instrumentation devices or software for testing or related engineering support services. The live testing usually takes place on government test ranges and facilities (which are rented by the contractors) under the supervision, or active participation of government employees (UK MoD 2005 XIX).

There are also problems with this arrangement, as contractors complain about delays because of understaffing and about facilities, which are not up to date in terms of technical standard. This led to the more recent tendency of some contractors having their own test ranges and facilities. In some cases contractors manage test facilities and conduct tests for the armed forces. For example, SAIC has its own Test and Analysis business unit which works together with the armed forces to evaluate the capabilities of combat and combat support systems. The company also holds a $500 million contract for supporting the Air Force's Operational Test and Evaluation Center. SAIC's activities range from scientific studies over developing instrumentation devices to large field-tests of new systems. SAIC has done this kind of work for the Air Force since 1989.

Increasingly prime contractors like Lockheed Martin, Northrop Grumman or Raytheon do test and evaluation of systems and components, often even of their own products, which raises some serious questions about the objectivity of these tests. Of course, there are practical reasons for this: testing is an integral part of the development phase and the contractor knows best which tests are needed; testing by the military research labs is time consuming, as it depends on available resources; leaving most of the testing to the contractors is in the end cheaper.

To shorten the development phase and to reduce costs, modelling and simulation is now used for virtually every test programme (Fox, Boito and Younossi 2004, XXI). The costs of live tests can be substantial. In respect to the testing of missile defence systems, which are on the one hand an extreme case, but on the other a weapons system in line with the general trend towards high-tech, a single test has come to cost $80 to $100 million (Graham 2003, 205). Alone the interceptor missile costs more than $35 million ($24 million for the kill vehicle and $11 million for the booster) (Graham 2003, 294). Naturally only few tests are conducted which generate the data used in simulations for performance assessments. Fewer tests also mean that less data is generated for analysis, which weakens the results of the analysis and also the accuracy of the models designed with the test data. As the reliance on simulations has grown, there is now the great danger pointed out by Wayne Biddle: 'Our weapons tests now use so much computer modelling and simulation that no one knows whether some new arms really work.' (quoted from Fox, Boito and Younossi 2004, xxi) As a result many of the most complex weapons systems have been fielded without sufficient testing and their actual performance fell far short of expectations. Out of a whole range of possible examples, three impressive failures shall be briefly discussed in this section: the Sgt. York Divad, the Patriot missile system and missile defence systems.

Sgt. York Divad A very illustrative example for the problems connected to testing is the Sgt. York Divad (Division Air Defense) automatic gun system, which was ordered by the US Army in 1978. The idea was to have an air defence gun, which would be guided by a Doppler-radar that could track targets and the gun would engage them automatically (quite similar to Phalanx). The only problem was that the gun could neither locate its targets, nor shoot straight. In a demonstration the gun turned towards the audience and some Army brass had to seek cover because the gun could not identify the aerial drone as a target, which was just in front of it. Despite this failure, the Army decided to continue the project and left the further testing in an incomprehensible act of negligence to the contractor. In another test Divad could destroy some aerial drones, but allegedly these were destroyed by a radio signal from the ground control station (quoted from Gray 1997, 60). The system was finally cancelled in 1986 after more than $1.8 billion had been spent.

Patriot missile The Lockheed Martin Patriot surface-to-air missile is a largely automated air defence system, which was hailed by the Pentagon after the 1991 Gulf War, when it was claimed that it was able to intercept the 43 and 38 Iraqi Scud missiles fired on Saudi Arabia and Israel with a hit ratio of 89 per cent and 44 per cent respectively. When challenged by the MIT professor Postol, who claimed that the missile could not be credited a single hit, the Pentagon admitted that the actual performance was worse than expected (Broad 2000). A GAO investigation indicated a nine per cent hit ratio overall (US GAO 1992). The system, which is now also in use by several US allies, has been fielded in the 2003 Iraq War, also with little success. According to a BBC investigation, the US Patriot system might have accidentally shot down a British Tornado killing the crew (BBC Online 2004). Apparently, the system malfunctioned hundreds of times during OIF and indicated there were incoming missiles where, in fact, there were none. The Patriot system operators had to use their (commercial) cell phones to ask soldiers in other near-by Patriot batteries whether the signals were real (Riggs 2004). The US Army simply blamed the Patriot crews and the RAF crew, which forgot to switch on its 'blue force tracker', and called it human error.

Missile defence Since the beginnings of SDI in 1983 more than $65 billion has been spent on some sort of missile defence capability and since the 1950s altogether almost twice as much ($120 billion) (Graham 2003, xxvi). There are still several active missile defence programs up and running, but few of them have met with expectations.[3] This hardly seems surprising, considering the problems with the Patriot system, which has targets that fly a lot slower than ICBMs (15,000 mph in mid-flight phase). There is no doubt that ballistic missile defence is the most ambitious technical military project in history. The complexity of it is mind-boggling. A system currently under development is the Terminal High Altitude Area Defense (THAAD) (also Lockheed Martin). It consistently failed in eight previous tests, until it miraculously shot down two missiles, which rescued the project from being cancelled by Congress. The surprising turn of luck looks suspicious, but the project will go ahead because already a lot of money has been invested and it is simply too important for prime contractor Lockheed Martin. In the meantime Boeing had not much luck with its ground-based midcourse missile defense system (GMD) either. The system uses an Orbital Science Corp. interceptor missile, which even failed to lift off during a test in December 2004 (Fulghum 2004, 34). Previous tests were inconclusive (five hits out of eight on easy targets). This did not lead to the eventual cancellation of the project, but rather to a $298 million contract for Boeing to upgrade the GMD system. Not only this, the US Air Force wants to go ahead with the programme and deployed the system in 2005 without further operational testing (Shorrock 2005, 10-14). General Kadish, the director of the Missile Defense Agency (MDA) defended this decision by saying 'You can't operationally test the system until you put it in place.' Critics on the other hand point out that without successful tests no one can know whether there is a capability for missile defence in a real attack (Shorrock 2005, 12). It has already become quite obvious that the initial goal of NMD of a near perfect interception capability has been abandoned when the Bush administration opted for a multi-layered system with reduced performance requirements (Graham 2003, 367). Another oddity about the deployment decision is that the distinction between testing and deployment has been effectively blurred. One wonders whether the operation of such a high-tech system can ever become routine and whether it has necessarily to remain indefinitely in an experimental stage.

This brief overview of flawed defence projects is not meant to suggest that the most modern weapons in Western inventories never work and are therefore useless, but rather to demonstrate that no one can be sure if, when and why they malfunction. In the case of new weaponry there is little or no operational experience available. Additionally, weapons are constantly upgraded to correct previous errors which occasionally introduce new and unknown errors. For this reason the Space Shuttle has rarely been modified because it is easier to deal with known problems than risking new unknown problems. Even after more than 20 years of experience with this system, catastrophes and near catastrophes (Challenger, Columbia and Discovery) indicate that spaceflights are still not routine. Very similar seem to be the problems in respect to high-tech weapons. Testing is always insufficient as the possibilities of errors are too numerous and can sometimes only be discovered once

3 The US service branches have their own missile defence projects running. The Air Force has the Airborne Laser (ABL), the Navy works on a missile defence upgrade for the Aegis system, and the Army has developed a missile defence system with ground-based interceptor kill vehicles.

they are 'beta-tested' in the field. Additional reason for concern is the growing involvement of private companies in weapons testing and evaluation activities, as this gives them better opportunities to rig tests and fake test results, which leads to the fielding of immature weaponry. This not only undermines national security and reduces military effectiveness when needed, but also obviously can lead to the loss of lives and in respect to nuclear weapons it can even have truly catastrophic consequences (see e.g. Perrow 1999).

Weapons Maintenance and Management

Modelling and simulation can be used to manage weapons stockpiles and to ensure that they are stored under optimal conditions and to determine the kind and extent of maintenance activities necessary for keeping the weapons operational. This application of M&S has its roots in the need to manage the nuclear weapons stock pile. However this technique is also used for many other weapons systems, which are not used or only rarely used and need to be stored for longer periods of time, e.g. missiles. Since the 1990s several weapons producers have found in life-cycle management of weapons systems a very lucrative market niche.

Nuclear stockpile stewardship and management The US and 180 other nations have signed the nuclear Non-Proliferation Treaty (NPT) and 44 nations have signed the Comprehensive Test Ban Treaty (CTBT) since 1996. These treaties prohibit the development, testing, and proliferation of nuclear weapons. In order to guarantee the US nuclear deterrent, President Clinton started the *Stockpile Stewardship and Management Program* (SS&M), also in 1996. This programme is intended to ensure the safety and reliability of US nuclear weapons in the long term and allow the US to retain its capability for building new nuclear weapons. About $5 billion annually are spent on SS&M and several major US defence companies are involved in it. The prime contractors are Lockheed Martin and Bechtel. SS&M has been called 'Manhattan 2' because of its tremendous size. Helen Caldicott has pointed out that the present US nuclear weapons programme is, in fact, bigger than during the Cold War (Caldicott 2002, 45).

Lockheed Martin has operated the Sandia National Labs in New Mexico since 1993 for $2.2 billion a year, where the company is mainly doing modelling and simulation of nuclear weapons explosions and degradation processes. This is needed not only for maintaining the present nuclear stockpile, but also for improving existing weapons and even for designing new ones. According to Helen Caldicott, the SS&M programme undermines the CTBT, as modelling and simulation combined with subcritical tests (nuclear tests, which do not involve a full detonation) enables the US to build new nuclear weapons such as the 'deep earth penetrator'. It also jeopardizes the reliability and safety of existing nuclear weapons, as the computer testing is hampered by an insufficient theoretical understanding of the bomb (Caldicott 2002, 47 and 58). There seems to be also a technical difficulty with accommodating new data derived from simulations and experiments with the older data from the times of nuclear testing, which makes the results questionable (Caldicott 2002, 47). The reliability of existing nuclear weapons could also be reduced by making unnecessary alterations to the warheads during ongoing refurbishments, as their original specifications are known and proven, while the modifications would be based on imperfect simulations with unknown outcomes (Kimball 2005).

After Sandia the other two of the three nuclear weapons design facilities in the US, Lawrence Livermore Labs and Los Alamos, are now ready to be taken over by private contractors. Lawrence Livermore, which used to be managed by the University of California, is also involved in the SS&M programme and conducts subcritical tests and simulations in support of SS&M. The University of California in alliance with Bechtel made a bid for Lawrence Livermore against Lockheed Martin with the University of Texas (Economist 2005a). Congress decided in December 2005 for now against Lockheed Martin. Finally, Los Alamos in New Mexico has been opened for a competition between Bechtel and Lockheed Martin with a contract worth up to $79 million annually, which was awarded to the Los Alamos National Security Consortium (University of California, Bechtel and Washington Group International) (Economist 2005b, 85-86).

The US nuclear arsenal does of course not only consist of warheads, but also of ICBMs, which are similarly complex and difficult to maintain. TRW (now Northrop Grumman Mission Systems) has been awarded a contract by the US Air Force to support and modify the entire fleet of 500 Minutemen III ICBMs until 2012 for up to $1.3 billion. The *Guidance Replacement Program* is supposed to extend the life time of the ICBM to 2020 and requires Northrop Grumman to do continuous tests of the ICBMs and their components, make software and hardware upgrades, and provide other technical support. Lockheed Martin together with Alliant Techsystems (ATK) is servicing the Trident I and II submarine based nuclear missiles. ATK has been recently awarded a $75 million contract by Lockheed Martin Space Systems to supply propulsion systems for the Trident II, which is planned for continuous deployment until 2042, while Lockheed Martin is the prime contractor for the Trident II life extension programme.

In the future the complete US nuclear weapons design and maintenance capability could be in private hands, which raises some serious questions. Should a nuclear weapons manufacturer (Lockheed Martin) be allowed to control all the nuclear research capability as well? Will the private management of nuclear weapons facilities increase safety and security risks? In any case, there are serious worries about the future of the nuclear stockpile because of the tremendous loss of knowledge caused by the retirement of the older generation of nuclear scientists. The other big worry is that the enthusiasm of private companies for unnecessary upgrades could have dangerous consequences, as this reduces, rather than increases the safety of nuclear weapons. Not to speak of the health and environmental risks involved in having such a big nuclear weapons programme up and running. It does not even serve a military purpose, as a tiny fraction of the present US nuclear stockpile would make a sufficient deterrent.

Nuclear weapons maintenance in Britain Britain has followed a similar approach to the US and has begun to privatize partially the maintenance and management of its nuclear arsenal. In 2000 the British MoD awarded a £2.2 billion contract to British Nuclear Fuels, Lockheed Martin, and Serco to operate facilities at Aldermaston and

Burghfield of the British Atomic Weapons Establishment for a period of ten years (in 2003 extended to 25 years). The AWE plc (as the joint venture is called) manages the research, design, maintenance, decommissioning, and disposal of British nuclear weapons. Its main responsibility is the management of the British nuclear deterrent, namely the 58 Trident II missiles based on four nuclear submarines. Most of the work of AWE plc is of a scientific nature such as developing tests for ensuring the reliability of explosives and trigger mechanisms, developing monitoring devices for detecting underground nuclear weapons tests, developing models and simulations of nuclear weapons performance, and developing counter measures against neutron and gamma radiation caused by the explosion of a hydrogen bomb. AWE plc was also tasked with the decommissioning of a nuclear reactor at Aldermaston and also takes care of nuclear waste and decommissioned nuclear warheads. Not long ago AWE plc claimed that it is not involved in the design of new nuclear weapons. Now the company is in charge of developing a replacement for Britain's Trident system. Serious questions remain.

British Nuclear Fuels Ltd. (BNFL) is a state-owned company, which operates nuclear power plants in the UK and also provides other related services to nuclear energy (e.g. nuclear reactor design, clean-ups) to the British government and international customers. Lockheed Martin does similar work for the US Department of Energy (see above) and Serco is an international facility managing company with military business in Britain, Canada, and Australia. Critics have pointed out that both BNFL and Lockheed Martin have some rather mixed track records in respect to nuclear safety. BNFL was charged in 2000 for poor safety equipment and measures and in 2005 the UK Atomic Energy Authority reported that 29kg of weapons grade plutonium had gone missing from the troubled Sellafield plant, enough for building four bombs or many more 'dirty bombs' (Gallagher 2005). The Y-12 plant at Oak Ridge, Tennessee, which was operated by Lockheed Martin Energy Systems, had to be shut down in 1994 because of poor management with no fewer than 1,300 breaches of safety regulations, some of which were critical (Mendick 2000). Serco is a company with no experience with managing nuclear facilities at all. From this perspective the decision of the British government to hand over the management of the AWE to these people looks like a rather strange move.

Non-nuclear weapons maintenance The US chemical weapons stockpile is managed by URS/EG&G, which is tasked with the destruction of these weapons as a result of the Chemical Weapons Convention of 1993. EG&G Defense Systems operates the US Army's Tooele Chemical Agents Disposal facility in Utah and has presently a $1.1 billion contract for building a similar chemical weapons disposal facility in Siberia including the training of Russian engineers to operate it (URS/EG&G 2004, 7). Similarly, General Atomics has been contracted to develop chemical weapons disposal facilities in Pueblo, Colorado, and Blue Grass, Kentucky (see www.ga.com). Up to now 167 states have ratified the CWC, and the destruction of the existing chemical weapons arsenals of the US, Russia, India and Libya will take many years. At the end of 2004 the US had destroyed 33 per cent of its stockpile, while Russia had only destroyed two per cent (SIPRI 2005, 613).

There are also many companies offering life-cycle support for conventional weaponry, which are mainly manufacturers, but also other technology companies. This will be discussed in the maintenance section of the 5th chapter. The basic idea of it is similar to nuclear weapons maintenance. The private companies continuously test weapons and their components, develop and run simulations in order to determine weapons degradation, schedule maintenance work, and regularly upgrade weapons systems.

The management, maintenance, testing, dismantling and disposal of nuclear and chemical weapons in the US, Russia, Britain, France and other countries will consume a tremendous amount of money and will go on for a very long time. According to estimates from the Brookings Institution, the overall costs connected to the US nuclear arsenal have amounted to $5.8 trillion in the 60 years of their existence (DeGroot 2004, 349). In the meantime new nuclear weapons are designed and none of the official nuclear powers has shown any willingness of completely giving up their nuclear deterrents.[4] The final disposal of these weapons will also be very expensive. Although it is doubtful that the responsible governments will make all the necessary investments, two things are quite obvious: 1) private companies like Lockheed Martin, Serco and BNFL will reap great profits from nuclear weapons related services in the coming decades; 2) we will continue to live with nuclear weapons in the foreseeable future, including the great health, safety and security threats posed by them. Many analysts even believe that the dangers of nuclear weapons seem to have increased rather than declined since the end of the Cold War. The privatization of their management does at least not inspire much trust.

Military Training

Computers have changed military training to such extent that the Pentagon speaks of a 'revolution in training affairs', which would be an important element of the current RMA (US DoD 2001b). As military equipment gets more sophisticated, more and better training is needed so that military personnel can effectively acquire and retain necessary skills. Training methods and devices become accordingly more important and need to be taken into consideration in the development of any advanced weapons. Military training simulations and simulators can be very complex and it requires a lot of skills and experience to develop them. There is now a close working relationship between the military and the industry in developing training aids and also in terms of carrying out military training.

NASA and the US military have used computer-based training since the 1970s.[5] Now all modern armed forces use computer-based military training at least in some fields, such as the training of pilots and tank crews. The possibilities for utilizing

4 Britain is presently considering building a replacement for its Trident missiles, which will reach the end of their life-times between 2020 and 2024. Although there is little public support for nuclear rearmament, the government seems determined to keep its nuclear deterrent (Oliver 2006).

5 The Tactical Warfare Evaluation and Analysis System at Las Flores on Camp Pendleton, CA. was built in 1979 and the simulator was based on 35mm slides.

computers for military training are nearly unlimited. Already all kinds of synthetic training aids are in use by the armed forces, or about to be developed. The advantages of simulations for military training are quite obvious. It allows saving on fuel and ammunition; causes no wear and tear on equipment; it allows to train difficult tasks without risk; it is easier to set up and requires less personnel than the use of the real equipment; scenarios, terrain, weather and other parameters can be easily changed and played through repeatedly; it is easier to monitor and analyse performance of the trainees; and finally, it enables military personnel to train at tasks, which cannot be trained in peacetime (Fisher 2005, 17). Additionally, more and more soldiers now belong to the 'video gaming generation' and are not only familiar with computer simulations, but also enjoy training on them (Joyce 2005, 16).

The Pentagon awards about $4 billion annually to private companies to conduct military training and run war games (Schwartz 2003, 100). Britain has also pioneered privatized military training with several innovative Private Finance Initiatives (PFI) (Campbell 2002). The British MoD has spent about £475 million on privatized military training through PFI in 2001-2002, most of it on flight training (UK MoD 2003c). In late 2006 the MoD has decided to practically completely outsource its training and to bring it all together under one contract called Defence Training Rationalisation (Pagnamenta 2006). This PFI is worth £10 billion and will include also the provision and operation of training facilities (including barracks, airfields and lecture halls). Many other NATO countries have followed the American and British example.

The main companies in the American military simulation business, which is truly huge, are Lockheed Martin, Boeing/McDonnell Douglas, Hughes, Evans and Sutherland, Raytheon, Northrop Grumman, CSC, EG&G, SAIC, Silicon Graphics, Reflectone, Viewpoint DataLabs, and Cubic. In Europe, the Canadian CAE, the German Atlas Elektronik (now a BAE SYSTEMS subsidiary) and the French Thales are particularly strong in the business.

There is usually a difference made between 'simulators', which are used to train individual soldiers on certain pieces of military equipment like vehicles or planes, and 'simulations', which can simulate any complex process, either for training or planning or the development and maintenance of weapons systems. This difference is not very clear-cut and even in the specialized literature the two terms are sometimes used interchangeably. A common typology is to distinguish between *simulator*, *virtual*, *constructive* and *live training*. Virtual refers to simulated computer environments, which are shared by multiple users; constructive refers to computer simulations at or above battalion level (Pryor 1999, 96) and live training is training on the real equipment.

Flight, Vehicle, and Dismounted Soldier Simulators

Since the development of the first flight simulators based on computer graphics in the early 1970s, private companies have become more involved in military training. The training equipment has simply become more sophisticated, which means more difficult to maintain and to operate. Support services related to computerized flight simulators were already common from early on. In the 1990s, the role of the providers of simulators was significantly expanded, as it was deemed more cost-efficient to let these companies build entire training facilities, which would be completely maintained and operated by them. Although military trainers do still decide on the curriculum and supervise the training, they now get extensive help from the contractors in developing curricula and conducting the training and post-simulation analysis. The following section deals with examples of privatized flight training in North America and Europe.

Flight simulator training Flight training was one of the training needs of the Western armed forces, which headed first for privatization for several reasons. General flight training has a civilian equivalent. There are many more pilots trained in civilian training programmes than in military ones and this goes back to the early days of flight. This means that the competence and experience with flight training is at least equal to, if not greater, in the civilian than in military aviation. Additionally, the training equipment was always designed and developed by private companies. Flight simulators actually go back to the late 1920s, when Ed Link, the father of flight simulation, patented a 'Pilot Maker' in 1929, which was a predecessor of modern day flight simulators. The Army Air Corps bought six Flight Trainers in 1934, which was the beginning of Link Aviation Devices, Inc. Today the company is named Link Simulation and Training Division under the parent company L-3 Communications. L-3 is now one of the main providers of military flight training in the US. For example, L-3 has already supplied eight of its Aviation Combined Arms Tactical Trainer (AVCATT) suites and also supports an US Army helicopter training programme called Flight School XXI. The company also trains pilots and operators on the E-3 Sentry AWACS, which includes ownership of the training devices and the training facility (L-3 Communications 2004).

Another example is CSC, which trains US Army helicopter pilots at Fort Rucker, Alabama, with the help of flight simulators and has an ongoing 12-year contract worth $1.1 billion. The contract requires CSC to 'provide, manage, operate, maintain and upgrade a suite of virtual flight simulators for Army helicopter flight training' (www.csc.com). EG&G is also involved in that project, which trains every US Army helicopter pilot and turns out 1,500 graduates a year (URS/EG&G 2004, 15). EG&G claims that 'More than 1,200 student pilots and 300 navigators are trained by EG&G annually for the US Air Force Air Education and Training Command at six Air Force bases across the country' (URS/EG&G 2004, 12). According to EG&G, the company has trained more than 20,000 US military pilots since 1989, in aviation theory, on simulators and on trainer aircraft (URS/EG&G 2004, 15). CAE has a training centre for C-130 Hercules pilots in Tampa, Florida, equipped with two simulators and eight multimedia classrooms, where it trains pilots from 20 different nations including US military personnel.

The British forces substantially privatized flight training in the late 1990s. A contractor maintained and operated training facility for Hawk jets was built in Anglesey in 1998. Serco has trained RAF helicopter pilots at the Medium Support Aircrew Training Facility at RAF Benson on Puma, Chinook and Merlin helicopters since 1997. The facility itself was built and is owned and operated by CAE. It includes six flight simulators and instructors, who conduct both computer-based and classroom tuition (www.serco.com). CAE claims that this arrangement 'has helped save the UK Ministry of Defence approximately 15-20 per cent compared to a traditional procurement' (CAE 2005). Other PFI include flight training for Apache attack helicopter, Tornado, and Eurofighter Typhoon aircrews. Boeing and AgustaWestland have formed the Aviation Training International Ltd., which supplies Apache aircrew training to the MoD on a 'pay-as-you-use' basis at four training centres in Sherborne, Dorset, for up to 30 years (www.agustawestland.com). One requirement was that the simulators are deployable in theatre, which could mean that contractors would have to deploy as well. Overall, this PFI seems to be a success (Strachan 2006, 17-19).

Thales holds a 30-year contract with the MoD under a £215 million PFI to provide training equipment and services at RAF Lossiemouth and RAF Marham for Tornado GR4 aircrews (UK MoD 2003a). Thales has taken over the management of most of the Royal Air Force simulators in 2002 (Penney 2002, 16). The company will also provide Eurofighter Typhoon training simulators and aids worth €200 million over a nine-year period. As mentioned before, for such new sophisticated military aircraft like the Eurofighter, training aids are not considered to be add-ons, but are actually developed in tandem with the aircraft. Even before the delivery of the first Eurofighter, a large European joint venture (comprising of Thales, Atlas and CAE Elektronik, Indra and Meteor) in form of the Eurofighter Simulation Systems (ESS) was set up in 1999. ESS is based in Munich and works together with the prime contractor for the Eurofighter aircrew synthetic training aid (ASTA) EADS. EADS has already begun to train Luftwaffe flight instructors in Manching on the very complex Eurofighter and will support ASTA together with ESS. Soon all British Armed Forces flight training will unified under the £2 billion Military Flying Training System contract.

In Germany military flight training is already supported by a single provider, as CAE maintains all of the German forces' flight simulators (Defense Week 2002, 11). It trains German Tornado aircrews at the German Air Force Flight Training centre at Holloman AFB in the US, which was built and is maintained and operated by CAE (www.cae.com). Another cooperation with the German forces includes the Helicopter Flight Training Services GmbH (HFTS), an international consortium consisting of CAE, Thales, RDE and Eurocopter, of which CAE holds 25 per cent. HFTS will design, build, and operate three centers in Bückeburg, Fassberg and Holzdorf and will provide NH90 helicopter training services to the German armed forces for a contract period of 2008 to 2022. CAE also has a joint venture with the Italian Agusta helicopter manufacturer with the name Rotorsim. In 2005 a new training center for A109 helicopters and other Agusta helicopters has opened in Sesto Calende, Italy. The training offered there includes flight and maintenance training, as well as classroom instruction and computer-based training (AgustaWestland 2004).

Even France has recently decided to outsource military private training and has awarded EADS a ten-year contract worth €175 million to manage and support pilot training at the flight school at Cognac. This is the first comprehensively outsourced function of the French military (Defense Industry Daily 2006).

Other Vehicle Simulators: Submarine, Ship, and Land Vehicle

Other pieces of sophisticated military equipment have also led to the development and use of expensive simulators and training facilities. One of the most remarkable privatization initiatives is the British outsourcing of training for nuclear submarine crews. In 2001 the MoD awarded a 30 year contracted to the FAST consortium worth £300 million to train crews for the new Astute class nuclear attack submarine. FAST comprises of CAE, BAE SYSTEMS, Finmecanica, and Flagship. These companies will take over different responsibilities in this project. The training center is in Faslane where the Astute class submarines are home-based (Defence Systems Daily 2001).

More recently BAE SYSTEMS won a ten-year contract worth £100 million to build training facilities and to conduct training (both simulation and classroom) for the crews of the new Type 45 destroyer, which is built by BAE SYSTEMS and which is about to enter service before the end of the decade. The facilities will be located at Collingwood, Portsmouth, Devonport, and Plymouth.

For surface ships embedded training simulators are used, which means that ship crews are trained with training devices that can be installed on the ship computer and targeting systems, which simulate real targets. The system in use by the US Navy is called Battle Force Tactical Trainer and it simulates what the ship sensors would show to the gunners while being involved in a real surface engagement. Its successor system *VAST* (Virtual At Sea Training) can even create virtual 3D environments of real terrain. The ship guns fire at a field of acoustic buoys, which determine where the round has entered the water and whether this represents a hit of the virtual target. Additionally, the system simulates the view a forward observer would have on the target, which is then used to adjust the gun and to direct the fire onto the virtual target (Chisholm 2002). The prime contractor is ACME Worldwide Enterprises, which does the systems integration and engineering, software development, test, installation, and evaluation.

In response to the challenges in Iraq, Lockheed Martin has recently developed a $9 million Virtual Convoy Simulator for HUMVEE crews, who have to drive through virtual cities, towns, villages and countryside, navigate, deal with ambushes, mines, road blocks, gun and RPG fire in an environment, which is also filled with civilians. The soldiers sit in a HUMVEE mock-up in front of a half circle shaped projection wall and they can fire weapons on any targets appearing on the virtual landscape. The simulator comes with a mission and scenario generator, a realistic virtual environment, and an after action review station.

There is already a great variety of tank and land vehicle simulators in use by the armed forces and the biggest providers are Lockheed Martin, Link, Atlas Elektronik and Oerlikon Contraves. First of all, there are driving simulators for the training of tank and truck drivers, who can learn the basics of tank or truck driving before they move on to the real thing. Then there are gunnery simulators for tank and artillery gunners, which are used to train gunners to quickly find or locate, engage and destroy all kinds of targets with the right ammunition. There are simulators for complete tank crews, in which the tank crew has to work together in order to achieve a given mission. And finally there are combat manoeuvre simulators in which several tank crews are linked together in one virtual simulation. Even if the simulator manufacturer does not operate the simulators or a training facility, it is completely normal that support services like logistics and maintenance, as well as instructor training are provided.

Dismounted Soldier Simulators

One of the greatest challenges for modern armed forces are military operations in urban terrain (MOUT) because soldiers are confronted with an immensely complex environment in which they are potentially exposed to enemy fire from '360 degrees'. Effective training for MOUT has become a priority for the US forces, especially since their Baghdad experience. US infantry soldiers now can get combat training in computer-based simulated training environments. Either by wearing a virtual reality headset, or standing in front of projected computer graphics, they can move around in the artificial environment of simulated cities, towns and villages, seek cover, and fire their weapons. As in real life they encounter friendly soldiers, unarmed civilians, and armed insurgents, and they operate in a squad and can call upon the assistance of computer-generated friendly forces. US PEO STRI operates the Dismounted Battlespace Battlelab at Fort Benning, Georgia, in which new simulation technologies for infantry soldiers are developed and tested. The realism of these simulations is amazing and while in cyberspace the soldiers cannot tell, which of the agents are computer generated or controlled by humans (Miller 2002).

Quantum 3D/CG^2 has been awarded a research contract to develop an embedded dismounted soldier training system, which will be man-wearable and consist of headmounted display and a head/body/weapon motion tracker and a training weapon. The next technological step ahead would be to get rid of the heavy virtual reality (VR) goggles and to project the computer image with a laser directly on the retina. This so-called Virtual Retinal Display (VRD) has been under development since the mid 1990s and can be used for VR based training, as well as to project information while the pilot/operator views the real world (Hambling 2005, 72). The technology is not yet fully developed, but in a few years virtual infantry soldier training could be part of the normal training experience. Other armed forces have also shown interest in virtual infantry soldier training, for example the Canadian Forces and the British Army.

A very low-cost version of the virtual infantry training experience is Marine Doom, the military version of the popular ego-shooter game Doom, which was developed for the Marine Corps Modelling and Simulation Office in 1997 (DerDerian 2001, 88). The monsters were replaced by human-looking enemies and Marines can train for combat in squads of four. The patch for transforming Doom into Marine Doom can be downloaded for free from the Internet. Marine Doom already looks very bleak from today's technological standards and in the near future reality and VR or mission and training will be practically indistinguishable with the only difference being that in VR one can get second chances, but in real life not. Anyhow, the US Army sees computer games such as Marine Doom not only as cheap training devices, but also as useful recruitment tools. In July 2004 the US Army released its first computer game America's Army: Operations, which can be downloaded for free and allows multiple players to train and fight as infantry soldiers in a sophisticated virtual environment with stunning computer graphics (Lenoir 2003, 175).

All these simulators increasingly utilize AI, which means that the supervisor of the training is no longer required to manipulate the myriad of parameters and elements in a scenario, sitting behind the trainee and giving feedback. The AI will automatically generate new scenarios and missions and will automatically analyse the performance of the trainee and will make suggestions for improvement. As the training becomes more automated fewer instructors and supervisors are needed. Ideally the training could be embedded in the used platform itself making it unnecessary to conduct the training at special facilities and making it possible to do it as a daily routine (Wilson 2004). Computerized training will also increasingly be conducted in the field. It is already the case that US troops in Iraq use mobile simulators in Iraq to train typical combat situations and for specific missions (Verton 2005, 28).

Virtual Training Simulations

In the 1970s military simulators were very limited in what they could do. They were stand-alone machines, which mainly allowed training a few difficult tasks on them such as landing on an aircraft carrier, and they were extremely expensive. In fact, the early flight and tank simulators cost roughly twice as much as the real equipment (Lenoir 2000, 308). Today simulators are quite flexible in terms of their programming and they can also be linked together for complex battle simulations in which a potentially large number of military personnel can take part. SIMNET, which was the first of its kind, is a simulator system network for M1 Abrams MBTs and M2/3 Bradley IFVs and was introduced into the US Army in the late 1980s. Since 1994 the US military has begun to develop JSIM, which has now replaced older virtual simulations of major war in Europe and includes a wide range of different scenarios and missions. JSIM allows joint training in a combined and distributed training event. All services will be able to train together on the same virtual battlefield. One component of the JSIM network is the *Close Combat Tactical Trainer* (CCTT), which was developed by Lockheed Martin for M1 Abrams and M2/3 Bradley armoured warfare training.

The same simulation system was recently bought by Britain. In the British version the CCTT consists of 170 vehicle simulators, which are linked together in a networked simulator suite and which allows the simultaneous training of up to 700 soldiers and commanders. The £250 million contract that has gone to Lockheed Martin includes the building and maintenance of two training facilities, one in Warminster, UK, and the other in Sennelager, Germany. Soldiers train in mock-ups of Challenger 2, Warrior and Scimitar tanks and armoured vehicles on 35,000 sq km of realistic virtual terrain, which uses real world terrain data. There is even the option to link other types of simulators to the CCTT network, for example helicopter simulators (Lockheed Martin 2002). In the future infantry soldiers can join the virtual battles as well and can practice joint operations with other service branches. Military commanders take part in the training, too and follow the virtual events from the Battlegroup Headquarters simulators. They do not only have to direct their forces in a live battle, but also manage logistics and organize repair and resupply for the units in the field. The training participants can either fight computer-generated forces or against human controlled forces in the second training facility.

Manoeuvres and Live Training Exercises

Simulators and simulations cannot, however, fully replace live training, which adds many more elements to the overall training experience. First of all, it takes place in real life by using the real equipment and, secondly, the soldiers have to endure physical stress and fatigue as they would have to during real operations. Realistic live training exercises are far more difficult to organize and often involve a lot of high-tech for simulating combat and weapons effects. There are several private companies, which specialize in organizing, supporting, and evaluating live training exercises, most importantly MPRI, Cubic, SAIC, Serco and Anteon. Below are a few examples of privatized military live training.

At the National Training Center at Fort Irwin, California, about 4,000 to 5,000 soldiers a month are trained in brigade size force-on-force live training exercises (www.globalsecurity.org). The facility and equipment is supported by several contractors and the training exercises are designed and conducted by Raytheon. In 1999 the company was awarded a contract worth $414 million to provide all kinds of technical services in support of NTC live training exercises, including instrumentation, logistics and engineering, software support, scoring, analysis and reviews, scenario development and battlefield effects (Raytheon 2002).

One of the first Combat Training Centers (Combat Maneuver Training Center Hohenfels) was established for the US forces in Hohenfels, Germany, by Cubic. On a large training area the US and other NATO forces can conduct simulated combat training, which involves training for armoured, artillery, and infantry units, and is based on the idea that two training groups fight each other while umpires watch the performance and intervene when necessary. Fire is simulated with the MILES system with which each vehicle and soldier has to be equipped and whenever a vehicle or soldier is hit with a laser it starts beeping, which indicates the kill. The American company SAIC has recently built the Egyptian National Training Center (also equipped with MILES) and helps with training analysis.

PEO STRI has awarded Anteon a five year contract worth $213 million for developing, building and operating a MOUT training facility, which will be used to prepare US soldiers for the challenges that lie ahead in the War on Terror. The company has now built the first MOUT training facility of the US military in Fort Polk, Louisiana, which consists of a 7 km MOUT training complex. Equipped with 1,000 video cameras and 350 microphones the training exercises are completely recorded and can be later analysed. The training facility is controlled by numerous computer workstations for instrumentation, control and management, and after action review. The results of the training exercises are afterwards presented in a 50 seat lecture theatre equipped with digital projection.

Since 1950 US forces have had a presence in South Korea and as the Korean War has never officially ended, they still maintain a high level of preparedness in face of a possible North Korean attack. Obviously US and South Korean troops have to train extensively and since 1998 this training has been in many respects privatized.

In Maehyang-ri Arctic Slope, which operates NORAD radars in Alaska, has taken over a bombing range from the US Air Force and conducts there bombing training on a small island, where A-10, F-15 and F-16 pilots practice bombing for up to 16 hours a day (Glantz 2003).

Blackwater USA specializes since 1997 in live training for military, law enforcement and security personnel at its own 6,000 acre training range in Camden, North Carolina. This is the largest private US military training facility. Blackwater offers a variety of courses there from basic shooting training over MOUT training to highly sophisticated Special Forces training, which shall prepare trainees to operate as security personnel in non-permissive environments like Afghanistan or Iraq (www.blackwaterusa.com). Blackwater also trains US Navy personnel at a 5,200 acre training facility in Moyock, North Carolina and has $35.7 million contract with the Pentagon to train 10,000 sailors a year in force protection (Yeoman 2003, 38).

Goose Bay, Canada, is a NATO training base where flight and weapons training is conducted. The RAF has awarded the British Serco a contract to operate the RAF's Southern Air Weapons Ranges at Goose Bay to provide all the support and training services related to it. Serco claims that it supports British aircrew training by 'covering targetry, air traffic control assistance and the operation and maintenance of range equipment' (Serco 1998).

Cubic has splendid experience in providing live training to armed forces, not only in the US, but also worldwide. Cubic is a prime contractor for 35 military training and training support facilities. On the company website Cubic claims to 'have designed and executed more than 6,000 simulation-based training events for US and allied forces throughout Europe, South and Central America, Asia and Canada' (www.cubic.com). Among other things, Cubic supports the US Army Joint Readiness Training Center at Fort Polk, Louisiana, where it has 1,200 personnel helping the Army with preparing scenarios, battle plans, background information and otherwise assisting the gameplay for the live training, e.g. creating battlefield special effects. Cubic has also run since 1991 the *Korea Battle Simulation Center* at Yongsan Garrison, Seoul, which conducts annual joint manoeuvres involving some 60,000 troops. Cubic designs and administers these exercises and provides other technical support services. Other duties include organizing training for US forces in Korea at the *Warrior and Counterfire Simulation Training Centers*, Camp Casey, for infantry and artillery training. Recently Cubic has helped organizing two major WMD training exercises in Qatar (Eagle Resolve 05) and Germany (Guardian Shield 05), which were used to determine the readiness of civilian and military authorities in dealing with WMD threats.

In a pilot project started in 2001, the German Ministry of Defence has for the first time contracted out military training to a private company, which has built and administered the new German Army Combat Training Center in Altmark, Sachsen-Anhalt for a three year period for altogether €75 million. This training facility consists of a control centre with 22 workstations, which are used to monitor, record, and supervize the live training with real tanks and vehicles. The tanks are equipped with a duel-simulator system (AGDUS), which is similar to the American MILES. Since January 2004 the training facility has been operated by the British Serco. The company provides all the technical support, including vehicle maintenance and technical services related to the training devices. Serco claims that up to now the feedback of the German Ministry of Defence has been positive (www.serco.com).

As the US Air Force and Navy no longer have enough pilots and equipment to conduct all the foreign military training that is offered by the US military to allies, a private company is trying to fill that gap. Advanced Training Systems International (ATSI) has bought 13 A-4 Skyhawk jets from Israel, has hired more than 100 pilots, instructors, maintenance personnel and administrators, and is located at the Williams Gateway Airport in Mesa, Arizona. ATSI has effectively created a small private air force and uses it for live training of foreign pilots (Scott 2003, 44-47). This includes so-called Red Flag training in which the trainees have to fight against adversaries. This is similar to the Red Flag training at Nellis AFB, Nevada, and Top Gun, Miramar, California. The company has already trained a first class of pilots from the United Arab Emirates in 2003 and expects a stable growth of demand in the future.

Organizing and administering such a sophisticated training exercise is beyond the capabilities of even technologically advanced armed forces such as the US military. The more other armed forces develop a taste for simulated training experience, the more they will have to involve private companies, which have the experience and technical capabilities to do it for them.

Finally, how effective are these training exercises? Very. At least this is the general sentiment in the US military. Walter Wriston tells the anecdote of a retired Army colonel asking an active Army colonel and Gulf War veteran for the reason of his great success during Desert Storm, despite the fact that none of his soldiers and officers had any combat experience. The other colonel answered: 'But we were experienced … We had fought such engagements six times before in complete battle simulations at the National Training Center and in Germany.' (Wriston 1997, 178) On the other hand, Chris Hables has made a good point by describing the real dilemma in respect to military training. 'The real problem is not simulator quality anyway, nor finding opponents to play with. The real problem is knowing what to practice in postmodern war.' (Gray 1997, 62) Obviously soldiers cannot practice every kind of the numerous different possible future missions and contingencies. At the moment nobody can say for sure what future warfare will look like. One thing at least seems to be certain: it will be diverse.

Combat Analysis and Wargaming

Developing new weapons systems brings not only the necessity to test them sufficiently in order to determine that they work under operational conditions, but also the need to determine how they should be deployed and tactically used. There is also the problem that weapons systems nowadays form larger networks and interact with other systems, which means that the introduction of a new weapons system has effects that go far beyond it. Again modelling and simulation can provide the analytical tools to develop and optimize new tactics and doctrine and to integrate new weapons into the framework of existing weapons systems.

In the industrial age the armed forces used huge sand tables and map boards for the planning and execution of military operations. In most less advanced armed forces this is still the case, but information age warfighting clearly requires much more sophisticated planning tools, which can model reality much more accurately and can take into account important factors like weather conditions, terrain, logistics, and so on. There are many think tanks, defence and IT companies, which specialize in modelling and simulation for combat analysis and wargaming. These simulations are sometimes used to shape operational military planning. The major players are the Santa Monica based think tank RAND Corporation, the Institute for Defense Analysis, Booz Allen Hamilton, Northrop Grumman/Information Technology (Logicon), SAIC, CACI, Anteon and BAE SYSTEMS/Marconi Integrated Systems.

Developing Scenarios and Wargames

Since at least the middle of the nineteenth century high-ranking military officers have been trained at staff colleges with war games for learning how to command troops in battle and how to develop war and battle plans. Little is known about war games outside defence establishments and the small circles of hobby war gamers, and it still is a rather arcane discipline. Many war games, especially those used for military planning, are secret and the available literature on the subject is fairly limited (Allen 1994, 9). There is also a problem in respect to the terminology. Many serious war gamers do not like the word 'game' at all and sometimes call it 'simulation', which is a very broad notion (Ghamari-Tabrizi 2005, 160).

There exists a great variety of different war games. Most of the war games now in use are computerized. Some of them do not involve any human players, while others are interactive computer games (humans playing against the computer), others are computer-assisted (humans playing against humans, but use computer modelling to determine outcomes of actions), and finally some of them are rather discussion and paper exercises.

The idea behind wargaming is very old and its modern version goes back to the sixteenthh century. A famous war game is the game called 'Kriegsspiel' or military chess, which was used by the Prussian general staff for officer training in the early nineteenth century. The game had complex rules and consisted of a map on which units could be moved and the outcomes of engagements with the enemy were determined by throwing a dice. William Frederick Lanchester developed a basic formula, which is in variation still used to determine the outcomes of engagements in war games. It basically says that 'The fighting strength of a force may be broadly defined as proportional to the square of its numerical strength multiplied by the fighting value of its individual units.' Or: $Nr^2 = Mb^2$ (Allen 1994, 67).

War games continue to be an important planning and training tool for staff officers and commanders. The involvement of civilians came with the emergence of the possibility of nuclear war and with the development of computers. Funded by the Air Force, the Santa Monica think tank RAND developed the first computerized war games in the mid-1950s to practice what has never been done before: fight a nuclear war. Nobody had naturally any experience in nuclear wars and there was no military tradition or history from which to draw upon. As a result, the civilians from RAND came into a position where they could tell the military how they should fight nuclear wars. Warfighting became then a mainly technical matter that demanded a rational and scientific approach and rested on quantification and computerized calculation, rather than intuition based on military experience and tradition (Ghamari-Tabrizi 2000, 163). The basic problem was how nuclear weapons should be deployed (countervalue or counterforce) and when and how they should be used (pre-emptively or in an escalation ladder). To solve this problem game theory and computer-assisted analysis was employed to find an optimum strategy for reducing the risk of a nuclear war and, if one has to be fought, to minimize the damage.

In principle, today's military computer simulations work similarly, but they allow a much more sophisticated and realistic simulation of military operations, as today's computers are powerful enough to calculate nearly all influencing factors. With the growing complexity of the computerized war games came a greater influence of civilian and private institutions and companies in designing, administering and conducting these war games in collaboration with the military. The first civilian-led war games were organized by RAND, but many other defence companies have quickly recognized the potentials and business opportunities connected to them. Vought's Missile and Advanced Programs division developed several war games to prove that NATO forces equipped with their new multiple rocket launch system (MLRS) can stop Warsaw Pact forces. This reduced the need for NATO to use nuclear weapons in the defence of Western Europe and encouraged NATO to develop MLRS (Allen 1994, 76). In other words, war games and simulations can be selling tools.

One of the companies involved in war game development, which rose fast to the inner circle of the US defence establishment, was BDM International. Between 1972 and 1992 the company grew at amazing 29 per cent or better a year and it developed the war games, which were supposed to prepare the US military to fight World War III. With the growing trouble of Communist insurgencies in the world in general and in Vietnam in particular, the Pentagon became interested in developing models and simulations for the analysis of revolutionary warfare. From this era originates the modern computer based war games, which had become as complex as war itself. The Douglas Aircraft Corporation was first to come up with a computer model of guerrilla warfare, when it was tasked in 1965 to game a possible communist invasion of Thailand (Allen 1994, 182). The model took into account the political-military dimension of guerrilla warfare and showed the importance of the control of population. The Vietnam War became a laboratory for getting the empirical data, which was needed for developing accurate models of guerrilla warfare and counter-insurgency strategy. The overall US strategy in Vietnam was heavily influenced by game theory, modelling, and wargaming. Most of it originated from the civilian, not the military world. After Vietnam there was another upsurge in computerized wargaming initiated by SDI. War games were used to show that the US could win a nuclear war against the Soviet Union in which the US could survive with minimum damage, if SDI systems could intercept 98 per cent of the Soviet ICBM in the boost stage (Allen 1994, 171).

However, war games can do more than just fight hypothetical battles. They are also powerful operational planning tools. Through computerized war games war plans for all kinds of contingencies and crises can be generated and kept on the shelf in case they were needed (Allen 1994, 3). By 1990 the Pentagon had an arsenal of no fewer than 363 war games ready (Gray 1997, 61). One of them was used for planning Desert Storm (OPLAN-1002-90). In fact the Iraq contingency was by 1990 one of the most wargamed scenarios at the time and the Army Concepts and Analysis Agency had already conducted a large wargame with the name Persian Tiger in 1989 resulting in the draft of OPLAN 1002-90 (FAS). BDM International, which was involved in Gulf War wargaming, allegedly sold a war game to Iraq in 1986 during the war with Iran. It modelled the Iranian transportation system and allowed Iraqi pilots to practice bombing runs on it before they went into real action (Allen 1994, 4). According to James DerDerian, another unidentified US firm had sold Iraq a war game or simulation, which it used for its operational planning for the occupation of Kuwait in August 1990 (DerDerian 2000, 15).

As can be seen in this brief overview, computerized war games have already been around for almost 50 years. What has changed during the 1990s is a much greater private sector involvement in the design and development of war games. Even back in the 1980s people like Andrew Marshall in the Pentagon's Office of Net Assessments had realized the great potential of commercial-off-the-shelf computer games for military training and planning. He invited James Dunnigan, a hobby war gamer and developer of commercial computer games, to the Pentagon because he was dissatisfied with the games he was getting from Pentagon game developers (Allen 1994, 93-95). Increasingly the Pentagon turned to private industry for war games. While some established think tanks and defence companies are still the main providers for simulations used in operational planning, combat analysis and battle-management, commercial products are increasingly used for general training purposes. There are obvious reasons for this: it takes the Pentagon many years to develop a new computer war game and many military wargamers simply do not want to wait that long, if they can have something that fulfils their training needs, that looks better and can be used intuitively, is readily available in a computer shop, and much cheaper than military designed games, they will make use of it.

Traditional War Games

Terrorism and the new WMD threat has become a favourite theme for contemporary war games. Booz Allen Hamilton has extensively wargamed WMD terrorism threats for the US government. The company, which holds most consulting contracts with the federal government and which produced revenues of $2.2 billion in 2002, has developed a war game for a bioterrorist attack and has written a threat analysis based on this game (Toigo 2003). The scenario was a chemical weapons attack on a stadium in Atlanta during a SARS outbreak in the city and about 85 government officials and experts participated in the game, which showed that the authorities would presently not be able to handle such a crisis (in the game the death toll was 50,000). Since 9/11 Booz Allen Hamilton has conducted over 100 similar games across the US for federal government agencies, local governments and also for the private industry (S. Harris 2004). Such war games are used to make decision-makers aware of threats, to find out about vulnerabilities to attack, to develop strategies and methods to counter them, and to allow decision-makers to get experience in mastering the most extraordinary situations.

Quite similar to the efforts of Booz Allen Hamilton is the work of Battelle in this field, a US technology company with an annual R&D budget of $2.9 billion. Battelle has developed an emergency plan for the Ohio state government that is supposed to enable it to operate in face of a major natural disaster or terrorist attack. For this project it has done an analysis of government functions and their interdependence and has prepared back-ups and logistics for an emergency. The company has also developed a virtual battlefield simulation for chemical and biological warfare for the US Army Maneuver Support Center, which is used to analyse and test the performance of presently available CB weapons detection and protection. Battelle advises the newly created Northern Command[6] in developing and shaping its mission and responsibilities, clarifying lines of authority, legal matters and strategies of interagency cooperation for homeland security. For this purpose, Battelle has developed scenarios and exercises for senior personnel (www.battelle.org).

Computerized War Games

JANUS was originally developed as war game for playing nuclear war, but it was later adapted for a theatre model of conventional war with tactical weapons. JANUS is still used by the US military in the National Simulation Centers and costs about $2 million a year in maintenance and operation. JANUS is also utilized by RAND to develop and test new concepts for future weapons and tactics. The war game can track up to 3,000 aircraft, tanks and vehicles on a 200 sq km size battlefield. War games and computer simulations are increasingly used to guide the way towards armed forces transformation and a lot of ideas and decisions derive from them. For this purpose the US military conducted a whole series of so-called battlefield experiments. One of them was the Force XXI programme at Fort Irwin, California. It was about testing a new smaller and lighter division equipped with digital communications. A lot of preparations work was done at the TRADOC Battle Laboratories between 1992 and 1994. In several advanced warfighting experiments (AWE) in Fort Bliss, Fort Leavenworth, Fort Knox, and Fort Polk, the US Army tried to develop its future organization and tactics. The Army After Next (AAN) programme that looks at the US Army in 2015 and beyond, is supported by many private companies, most importantly by MPRI.

In 2002 the Pentagon tested its joint warfighting capabilities in an experiment called Millennium Challenge 2002, which cost the impressive sum of $250 million. The goal was to better understand the lessons learned in Afghanistan and the exercise was a combination of tabletop game, virtual war game and live exercise. It took place at several locations, among them Fort Irwin (ground operations), Nellis AFB (air operations) and off the coast of California (naval operations). According to a participant who commanded the opposing forces, the experiment was ‘fixed’ to let

6 The US military is organized in five regional and five functional commands with numerous subordinate commands. Northern Command was established in 2002 and is responsible for the defence of and military operations within the North American continent. Its main functions are homeland defence (aerospace, land and sea defence) and support to civil authorities in times of national emergencies.

the 'blue' forces win, rendering the results rather dubious, as in real life you cannot simply order your enemy to behave as you wish (Doborek 2002).

The US Space Command has recently conducted a big space war game at Schriever, AFB, Colorado, called Schriever III (Schriever I and II took place 2001 and 2003), which involved 350 professionals from 20 agencies from the US, Canada and the UK, in February 2005. The game highlighted the importance of space systems for expeditionary forces in regional conflicts like Afghanistan and Iraq and also their importance for battling terrorism. The details of the game are still classified, but it was a future scenario playing in 2020. The game was based on a simulation developed by Northrop Grumman with the name Missile Defense War Game and Analysis Resource (MDWAR). Northrop Grumman has a partnership with the Joint National Integration Center (JNIC) at Schriever AFB and conducts about 15 war games a year to develop and test missile defence systems for JNIC. Of course, Northrop Grumman does more than just develop scenarios and war games for missile defence, it is also involved in two missile defence programmes, the Kinetic Energy Interceptor and the Airborne Laser programmes (Northrop Grumman 2005). The first Schriever space war game conducted in 2001 cost about $1.5 million, according to Air Force cost sheets (J. Singer 2001). Northrop Grumman is teamed at JNIC with Raytheon and SPARTA Inc. and has 300 contractor and 500 subcontractor personnel at Schriever AFB (Northrop Grumman 2005).

Commercial War Games

Commercial war and strategy games like *Civilization* or *Red Alert* are immensely popular and have influenced the design of military games as well. However, they will not easily make it into the military world in form of one-to-one adaptions, as the military acquisition process is very complicated and really favours the established suppliers. It is therefore more likely that defence companies will take advantage of the talent and experience available in the commercial sector for developing their own war games. There is at the moment already a clash of cultures occurring between military versus commercial wargaming. Military war games are based on real world data of the performance of weapons systems and are basically attritional models in which the bigger guns win. In commercial wargames other factors can be decisive as well such as morale and fatigue.

The criticism of the military game developers is that it is completely arbitrary to give military units values for morale and fatigue, as these cannot be measured or quantified. Therefore, the military game developers claim to base their simulations on much more accurate models, while commercial games have arbitrary and inaccurate models rendering their results invalid (Peck 2003). Critics say that the military models are inaccurate too, as the data on weapons performance is often based on laboratory tests under optimum conditions and not on tests under operational conditions, which makes these weapons systems look much more effective in a game than they actually are in the real world. For example, weapons tests of the Sidewinder air-to-air missile indicated a hit ratio of 98 per cent under optimal conditions, which was the value used in simulations, while the real world performance of the Sidewinder in the Falklands campaign only showed a hit ratio of 73 per cent (Allen 1994, 290). An additional problem is that, according to James Dunnigan, the data needed for accurate modelling sometimes does not exist or is scattered amongst the various military labs and not easily available. As a result 'a lot of times, when a number is needed, it is made up' (Freedberg 2001). As the models on which war games are based become ever more complex, it also becomes extremely difficult to validate and verify the simulations, as nobody really understands what is going on inside them (Freedberg 2001). It is sometimes hard to tell whether some simulations have any analytical value at all, as there is not enough empirical data to either verify or disprove them. The great danger is that computer simulations can look extremely convincing to even a specialist. A participant of a RAND war game noted that 'Perhaps the biggest problem that exists today is that a war game model is extremely difficult to understand. Years are spent in their development, and when someone is in a hurry for an analysis, he can't spend the time to learn all vital details of the model.' (Ghamari-Tabrizi 2005, 172) Whether it works one might only find out after billions of dollars have been wasted in useless weapons systems.

Operational Planning and Lessons Learned

Private companies, institutions and think tanks increasingly get involved in matters of defence analysis and operational planning. This is a very recent development that has begun only about ten years ago. As the number of threats and contingencies has tremendously increased and also the technology needed to address these threats, there is now the tendency in particular in the American and British armed forces, to simply contract out some of the analysis and planning work or at least aspects of it, which is needed for developing contingency plans or for understanding the 'lessons learned'. All the private sector involvement in military planning has either to do with providing technical planning systems, intelligence support, or logistics planning. Many of the big defence companies design mission planning tools, which help the military to organize the logistical requirements for particular operations, e.g. air operations. For example, SAIC has developed 2,500 mission planning and support systems for a wide range of military aircraft, including the F/A-18, AV-8B, F-111, MH-60, MH-53E, AH-1, KC-130R, KC-135 and V-22. These were sold to Argentina, Australia, Finland, Italy, Malaysia, Spain, Switzerland and the US forces (Ebutt 2004, 116). They are mainly logistics planning systems that are quite important for operational planning as well. The logistics side will be dealt more extensively in the section on logistics in the next chapter.

Once the action is over there needs to be an analysis of what went wrong and why. After Operation Desert Storm in 1991 a great number of veterans complained about health problems, which they attributed to their participation in the war. 'Gulf War syndrome' was (and still is) a mystery. BDM International was hired to investigate the matter and the company conducted a large survey, questioned veterans, documented the cases, collected data and made databases, analysed troop movements and locations, and developed analytical tools. The contracting agency (OSAGWI) was later criticized by GAO for improper contracting practice: the contract was awarded non-competitively and did not clearly specify what BDM was supposed to do. However, OSAGWI has pointed out that the contract with the BDM was to expire in 2000 and 'that the need for the type of support services that BDM is providing will continue for an indefinite period' (US GAO 2000).

The clear-up of Operation Iraqi Freedom was also contracted out. The think tank Institute for Defense Analysis (IDA) was tasked to analyse the lessons learned from the campaign OIF. IDA is a semi-governmental organization, being an independent non-profit organization working exclusively for the US federal government. Different to any other war conducted by the US forces, in the Iraq War the 'lessons learned' analysis was not an afterthought, but was happening right from the beginning of the operations. Analysis teams were present for observation and collection of information while the war was unfolding. The Joint Forces Command set up a 50 people strong analysis team, as part of the Joint Advanced Warfighting Program (JAWP) headed by IDA, which analysed the war from posts in theatre, in Europe and the US. Their job was to highlight possibilities for improvement in performance and to make recommendations for changes. IDA has also extensively analysed the 1991 Gulf War and has developed a sophisticated computer simulation of the tank battle 73 Easting, which took place on February 26, 1991. Developing the simulation involved determining exactly all the engagements that took place, including when, where and with what result weapons were used. The battle could be replayed repeatedly with different parameters. For example, it could be assessed what kind of impact it would have had, if the Iraqis had nightvision equipment (Lenoir 2003, 177). Such kind of after-action analysis can be very useful for planning and executing future battles.

Apart from IDA, another private think tank, DFI International was contracted by the US government to do military operational analysis of the Iraq War. DFI International assessed the effectiveness of strike aircraft in Afghanistan and Iraq. The company also did an analysis of force structure requirements of future capabilities and operations for the US Air Force. DFI started in 1984 mainly as a business intelligence company and expanded in 1994 into the defence and government sector and has now 170 full-time employees (www.dfigov.com).

Similar analysis work has also been done by the much bigger MPRI (700 full-time employees and 10,000 'on call') in the aftermath of the 1991 Gulf War. MPRI later also presented its findings, lessons learned and recommendations in a series of briefings to senior officers of Sweden and Taiwan (Lanning 2005, 198). Some other analysis conducted by IDA consisted of several studies concerning the Strategic Agility and Adaptability, Bandwidth Utilization, and Enabling Capability (www.ida.org). IDA has also developed the Contingency Operations Support Tools (COST), which is a simulation used for calculating the overall costs of contingency operations.

Altogether it can be concluded that direct private involvement in military planning is still very limited and mainly concerns the provision and servicing of technical systems, logistics planning, and analysis work. On the other hand there are some indications that private companies could become more influential in military planning, as the designers of new high-tech weapons know best their limits and might understand better how to effectively employ them in combat. Attempts of prime contractors such as Northrop Grumman to get into this area of developing new operational concepts and operational planning with its Analysis Center illustrate this trend.

Developing Military Doctrine

Private companies are not only involved in the purely technical side of military training, but also in the education of officers and the formulation of doctrine. Because war has become so technical, it has opened the field to civilian analysts, who are often better in analysing the technical and managerial side of war than the military itself. Although activities such as teaching at staff colleges and writing field manuals are not technical in nature, they derive directly from the technical aspects of war and the technical competence civilians can have.

There are not many companies, which offer such staff services, as this is one field where not much outsourcing has taken place yet. However, there is a trend towards outsourcing less technical services and it is quite likely that there could be much more outsourcing of military education and doctrine development. The dominant company in this field, even from a global perspective, is clearly MPRI, which is quite unique in its business concept. The company basically sells military expertise and claims that it can draw upon the knowledge and experience of over a thousand former US military personnel, among them many retired generals, admirals and staff officers. MPRI became famous after its support to the Croatian army, which allegedly helped the Croatians to stage the two successful offensives *Storm* and *Flash* in 1995. This story is well known and has been repeatedly analysed (Singer 2003, 123-130; Avant 2005, 101-113; Silverstein 2000, 171-174; Lanning 2005, 196-204). What is less known is MPRI's work for the US military, which is, in fact, quite extensive and includes a wide range of consultancy, training and recruitment services. MPRI has run part of the Reserve Officer Training Corps programme and trained military officers at 217 universities between 1997/98 and 2002. Then the contract was awarded to COMTek after a rebid (Avant 2005, 118). According to Linda Robinson, 'MPRI and Cubic are developing curricula, analyzing and writing doctrine at numerous military schools around the country, conducting general officer executive education, and training press officers' (Robinson 2002). The company maintains offices at many US service schools including, Fort Leavenworth, Kansas, Fort Knox, Kentucky, Fort Still, Oklahoma and Fort Bliss, Texas (Lanning 2005, 201). MPRI runs extensive training programs for the new NATO members in Eastern Europe. For example, MPRI helped to modernize the Croatian and Bosnian armies and does now the same for Bulgaria (Lanning 2005, 201). It has recently opened a simulation centre for the Romanian armed forces in 2002, where it trains Romanian officers in NATO tactics and helps to ensure the interoperability of Romanian equipment and doctrine with NATO standards and requirements.

Even in Britain private companies are involved in military education and doctrine development. The British Serco boasts: 'We led a consortium to design, build, finance and operate the Joint Services Command and Staff College in Wiltshire – one of our many UK Ministry of Defence contracts.' (www.serco.com) The JSCSC was founded in 1997 as an educational institution for all British services, as well as foreign armed forces and civilian personnel. Serco was contracted to manage the facilities including all the technical equipment for a 30 year period, while the teaching was contracted out to King's College London (Monahan 2001). Serco was also recently awarded a contract in 2004 to develop a tri-services military training college in Muscat, Oman, in cooperation with the local Bahwan Engineering Group and BNP Paribas. The contract is worth $165 million and the facilities are planned to be ready after 18 months of the start of the project. Under a PFI Serco will operate the facility and train Oman's armed forces for a period of 30 years (MEED 2004). Similarly, Booz Allen Hamilton runs the Saudi Armed Forces Staff College, which not only includes tactical training, but also teaching senior-level military skills (Silverstein 2000, 181).

Sometimes private companies are tasked to write military handbooks and to develop military standard operating procedures. For example, MPRI has written a field manual on how the US Army uses contractors in contingency operations (FM-3 100.21/ Contractors on the Battlefield). MPRI is not alone in the business of writing military doctrine. There is, for example, LOGCAP. The Pentagon opened a competition for a contract to write a study on how a private company could provide logistical support to the Army in 13 different locations around the world in 1992. KBR won the contract, which was later expanded to plan for other contingencies as well, and the result of KBR's effort was the Logistics Civilian Augmentation Program (LOGCAP). This is a logistical contingency plan that details how a single contractor could manage logistics for a wide range of contingency operations. This KBR plan became the basis for the LOGCAP contract, which was held by the same company between 1992 and 1997 and from 2002 onwards (a ten year contract this time) (Briody 2004, 184). Another example is Battelle, which has written a technical field manual for the use and maintenance of the Joint Biological Detection System. The field manual instructs military personnel how to use biological weapons detection devices. As the system is, according to the company, quite complex, it represented a serious challenge to make it understandable and usable for the ordinary soldier without any technical education.

Defence think tanks like RAND or IDA, particularly in the US, have substantial influence on the development of strategy and doctrine. They develop new concepts and ideas for the development of new weapons, the improvement of doctrine, and the employment of particular weapons and other systems. This development clearly indicates a growing uncertainty of the military in respect of its own competence and, to some extent, a shift of expertise and authority for military affairs from the military professionals to the private sector and its civilian defence analysts.

Conclusion

Computer simulations are quite universal tools which can be used in many areas of the commercial and military world. Simulations are one of the key technologies that will shape a future high-tech military. Already today this technology impacts on weapons development and maintenance, military training and military planning and doctrine development.

Computer simulations have become essential for the development and maintenance of weapons systems, especially nuclear weapons. Much of the responsibility for weapons testing and life-cycle management has shifted to the private sector. This is on the one hand a very convenient arrangement for the armed forces. On the other hand, it is a rather worrying development. Privatized weapons tests might not be very objective and the use of too much simulation for weapons development might mean that their actual performance characteristics are not accurately determined. The tendency of continuously upgrading equipment exacerbates the problem of having lots of immature and unproven weapons.

The Pentagon views the computerization as a 'revolution in military training' and has invested heavily in it since the 1970s. With the computerization of flight and later tank crew training, slowly the privatization of military training began. Now privatized military training is big business and still a growth market. A mixture of a rather limited number of companies has come to dominate this market. Most of them are manufacturers of simulation equipment. These companies operate globally and are heavily involved in foreign military training programmes. They also help to organize large training manoeuvres, which are sometimes international with soldiers from many armed forces participating. At the moment complete military capabilities are not really for sale, but certainly the training and expertise to acquire them.

The computerization of war games has also led to a greater private sector involvement in operational planning and doctrine development. The use of sophisticated war games for analysis and planning purposes is still most common in the US, but other armed forces are trying to catch up. Thanks to the international activities of companies like Booz Allen Hamilton or MPRI many US allies can take advantage of their expertise in conducting war games. The current trend indicates that simulations will be increasingly used for operational planning and for guiding the transformation of the modern armed forces. They are needed to deal with the increasing complexity of technology and the complexity of logistical requirements that derives from it. Not too far into the future all major military operations will have been extensively wargamed before they take place and armed forces will worry about encountering situations that are 'off simulation'. In the end, war games are nothing short of a weapon and private companies are those who wield this weapon.

Chapter 5

Operational Support Services

Contractors play an increasingly important role in sustaining and providing military capabilities in support of contingency operations. This chapter looks at contractor activities in connection to maintenance and field services, technical services related to modern logistics, and at cases where contractors operate equipment and systems for the military. When it comes to the American and the British forces, it is quite obvious that they simply cannot operate in any theatre without contractor support. It is less known that other armed forces are also becoming more and more contractor dependent, at least with respect to certain technical services in connection to lift capacities, base operations and other logistics services. In international contingency operations, e.g. in the Balkans and Afghanistan, such armed forces like the French and German, which are notoriously reluctant to use contractors in deployed operations, have used them in at least some functions.

Maintenance, Overhaul and Upgrading

For reasons outlined in chapter 2, the defence industry plays now a crucial role in maintaining some of the most sophisticated weapons systems. There are two types of contractor systems support: *Interim Contractor Support*, which is only needed during the introduction of a new system, and *Contractor Logistics Support*, which is a more permanent arrangement for the use of contractors to support particular pieces of equipment. Both types of contractor support activities have increased over the last 15 years. As a result of force reductions and low retention rates, the armed forces are short of qualified equipment maintainers, who are needed for keeping aircraft, vehicle and other equipment operational. The situation is made worse by the fact that the rather boring and frustrating maintenance work does not appeal much to prospective soldiers. The military is therefore happy to get rid of these tedious duties, which affect recruitment and retention rates (Spearin 2003, 29). The most sophisticated and rare equipment (e.g. combat aircraft or UAVs), or some important components of it, tend to be maintained by the original equipment manufacturers. When it comes to new equipment, which is rushed into battle, there is simply no other option than to rely on contractors, as military personnel are simply not trained on it (Taylor 2004, 192). Less sophisticated and more common equipment, especially such which has civilian equivalents (e.g. transport aircraft, land vehicles), is sometimes maintained by companies that specialize in technical services, like Raytheon, EG&G or DynCorp. Whenever certain equipment is deployed in a theatre there is a high probability that at least some contractors have to be deployed for equipment maintenance.

There are different types of technical equipment support activities, which need to be distinguished. First, there are basic maintenance activities, which need to be done on a weekly or even daily basis. Secondly, there are repair activities for damaged equipment, which is infrequent and only done when needed. Thirdly, there are overhaul activities, which need to be performed at least every couple of years for repairing or exchanging major components. And finally, there are upgrading activities, which modify existing equipment in order to increase its performance and/or capability. It is common that the basic maintenance and repair tasks are performed by the armed forces themselves, while more difficult tasks such as overhaul and upgrading is left to contractors, in many cases the original equipment manufacturers. Sometimes overhaul and upgrading has to be done in the field, as it presently happens or has happened in respect to some US equipment in Iraq.

Contractor-provided equipment support seems to be a universal phenomenon and it is not limited to the usual suspects, the US and UK. It has also become (to a more limited extent) quite common in other NATO countries like Germany, Italy, Spain and Netherlands. It can be said that the US and UK are much less reluctant to use contractors for equipment maintenance in forward deployments than the armed forces of other countries. For example, Germany has begun to privatize parts of the weapons and equipment maintenance within Germany, but does not yet want to use contractors in deployed operations.

The US on the other hand, is at the other end of the spectrum and relies so heavily on contractors in contingency operations that even the US General Accounting Office was not able to determine the full extent of contractor support provided to US forces worldwide, as neither DoD or the services know it (US GAO 2003, 1). The DoD spends about $59 billion annually on the maintenance of its 30,000 combat vehicles, 15,000 aircraft, 900 strategic missiles and 300 ships (US DoD 2004a). About 46 per cent of these weapons systems are maintained by contractors and it is planned that in the future the figure will rise to 50 per cent (Blizzard 2004), which is the limit set by Congress.

It is quite normal that there is contractor support available in the introduction phase of a new weapon, which used to be just for the first year. Congress now requires four years of contractor support. For non-critical weapons systems this can be extended to their whole service life (Castillo 2000). In most cases maintenance is performed in collaboration with industry and in a few cases the US military has no own maintenance capability at all (C21J executive aircraft and Guardrail). In other cases the support capability is very limited. For example, the National Guard has no independent support capability for its Apache and Blackhawk helicopters. When National Guard units were sent to Afghanistan, they had to rely on contractor support for their helicopters (Taylor 2004, 191). According to a RAND study on contractor support in the recent Iraq War, 45 weapons systems and 27 aircraft types relied on contractor logistics support of which 42 received more than 75 per cent of its maintenance from contractors (Camm and Greenfield 2004, 3). This indicates that contractor support is far more crucial for US expeditionary forces than the lower figure of 46 per cent indicates.

Britain has followed the American example and has already outsourced a large proportion of equipment and weapons maintenance. In 1999 the MoD set up the tri-service logistics provider Defence Logistics Organization (DLO), which provides and directs logistics for the British armed forces worldwide. As pointed out by Trevor Taylor, the DLO has moved away from the provider function towards a mere 'decider' function in respect to equipment support (Taylor 2004, 184). The DLO's plan for 2005 explicitly states the goal that it wants to utilize industry capabilities to a greater extent in cases where there is an increase of effectiveness and no additional operational risks and costs. The DLO also anticipates that in the future contracting will move up the transformational staircase from traditional contracting to spares inclusive contracting to availability and finally capability contracting (UK MoD 2005, 14).

Table 5.1 Transformational Staircase (UK Defence Logistics Organisation)

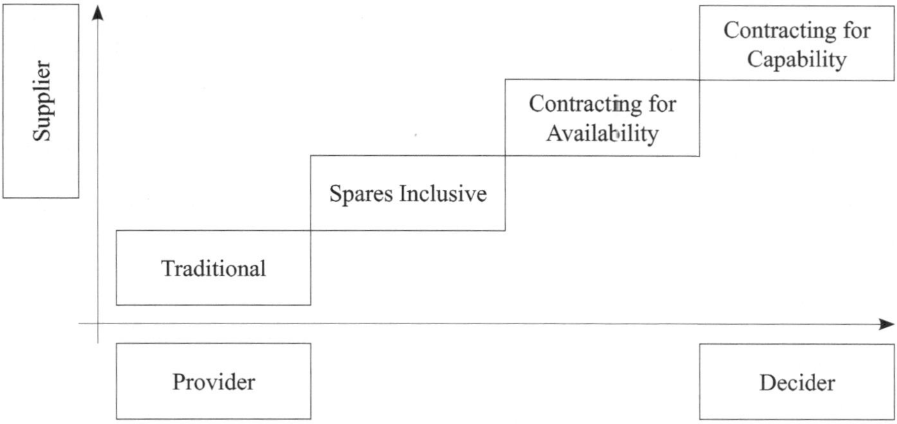

Source: UK MoD (2005), *Defence Logistics Organisation: Plan 2005*, p. 13.

Effectively, it means a much greater dependence on industry and civilian contractors, which can already be seen in the recent Iraq War, which required the deployment of no fewer than 1,500 civilian contractors. They were needed because there were not enough qualified military personnel to maintain and repair the most sophisticated weapons (UK MoD 2003, Chapter VIII). For comparison, the 1991 Gulf War only required the deployment of 100 civilian contractors (Taylor 2004, 195).

Canada has a similar approach like Britain and gives more responsibility to industry for maintaining and repairing weapons and equipment. The Defence Planning and Management organization has started a range of new capability programs, which are intended to improve the sustainability of the Canadian Defence Forces at home and overseas. The Optimized Weapons System Management (OWSM) initiative aims to streamline organic equipment support and to determine life-cycle support requirements.

The German armed forces have slowly begun to reduce organic maintenance capabilities and to contract out maintenance for some of its equipment. According to a report of the German Federal Accounting Office, about six per cent of equipment maintenance of the German Army has been outsourced so far (Bundesrechnungshof 2003, 165). The German Army deemed it more cost effective to rely more on industry than to optimize its own internal capabilities. The outsourced maintenance was largely a high quality overhaul of Army vehicles and equipment in order to increase the equipment readiness from presently 60 per cent to 80 per cent. Should cost-savings materialize then more maintenance work will be outsourced in the future.

Aircraft Maintenance and Field Services

The aircraft industry was first to push for the privatization of maintenance services for military aircraft. This practice goes back to even the 1950s with the beginning of the nuclear arms race (Markusen 2001). Modern combat aircraft are notoriously maintenance intensive items. Mary Kaldor writes that for every flight hour of the F-111 bomber and the F-14A fighter jet 98 hours of maintenance is required, for the F-4J it is 83, for the F-15 it is 34 and for the least complex A-10 it is still 18 hours (all of the aircraft are still in use) (Kaldor 1983, 132).

United States Private contractors maintained US aircraft in Korea and Vietnam. During the Vietnam War a lot of dual-use aircraft like Huey helicopters and transport aircraft (C-123, C-141) were maintained by contractors, most of them by DynCorp, which had 3,000 maintenance workers in Vietnam (Schwartz 2003). During the 1991 Gulf War most of the about 5,000 contractors, who deployed in the Gulf region, maintained and repaired aircraft and components.

Today the Pentagon has only a limited maintenance and repair capability for some of its aircraft, especially for some which are considered critical, such as the F-117A and the B-2 Spirit (US GAO 2001). Especially with respect to aircraft maintenance, the DoD makes a lot of use of Contractor Logistics Support arrangements, where manufacturers agree to provide spares, logistics and technical services required for keeping the equipment operational in the field. For deployed operations the Pentagon also uses so-called Contractor Field Teams or Contractor Field Services (CFS), which are presently provided by just a few companies: Raytheon Technical Services, EG&G/Lear Siegler Inc. (LSI), DynCorp, L-3 Communications/Vertex Aerospace, Booz Allen Hamilton (teamed with Brown and Root Services) and Lockheed Martin. These companies have agreed to provide technical services on aircraft and other specified equipment to forward deployed forces. However, it is quite difficult to find information containing any details on locations, numbers of employees, or activities, as the contracts under which contractors deploy often do not explicitly mention support for deployed operations (US GAO 2003, 27) and the companies concerned only give very general information on their activities.

Examples for contractor support for deployed operations are: L-3 Communications has several contracts with the US Air Force to maintain, overhaul, and upgrade many of its aircraft within the US and worldwide, including the B-52, the B-2, the A-10, the F-16, and C-130. For the US Army it presently provides support to armoured vehicles, and the AH-64 Apache, UH-60 Blackhawk, C-47 Chinook, and OH-58 Kiowa helicopters deployed in Iraq and Afghanistan. DynCorp has supported the

US forces in Korea for more than 50 years and has still more than 200 employees in Korea and Japan, who service AH-64 and UH-60 helicopters. EG&G/LSI claim to service US military aircraft and vehicles in 80 locations in the US and abroad, among them locations in the Balkans.

Lockheed Martin supplies a whole range of military aircraft to the US DoD and other air forces around the world, among them the A-10, the U-2, the C-130, the S-3 Viking, the F-16, the P-3 Orion, the F-117, the F/A-22, and the JSF. The company also plays a major role in maintaining and modifying these aircraft for its customers. For this purpose it has formed Lockheed Martin Aircraft & Logistics Centers (LMALC), which consists of several technical service companies that were unified under LMALC in 1997 (Hoover Online). The support includes work done on aircraft, air frames, engines, and related logistics services, also overseas. LMALC claims to have presently 80 Contractor Field Teams in support of deployed operations worldwide. Its most important customer is the US Air Force and between 1997 and 2002 the LMALCs have produced $323 million in revenues (Center of Public Integrity). EG&G claims that it provides maintenance services for virtually every aircraft in the US inventory and that it overhauls and repairs jet engines of some US aircraft types. It has overhauled and upgraded every new UH-60 Blackhawk helicopter in the US Army inventory and has modified hundreds of C-130 aircraft with night vision and weather radars (URS/EG&G 2004, 15). The US Army has outsourced the maintenance support for its AH-64 Apache helicopters to Team Apache Systems (a joint venture between Boeing and Lockheed Martin). This includes life-cycle support, logistics management and field support for the Apache helicopters. The US Army made a similar arrangement in respect to its M109 Paladin self-propelled howitzer (Drake 1999).

Britain In 1999 the British MoD combined RAF and RN aircraft maintenance into one joint organization with the name Defence Aviation and Repair Agency (DARA). DARA manages all MoD fixed-wing and rotary-wing assets, including engines, components and electronics and is Europe's largest government-owned aircraft overhaul, repair, and maintenance facility in four locations (St Athan – fixed-wing, Fleetlands – rotary and engines, Sealand – electronics, and Almondbank – components). It has also commercial business and does repairs on civilian aircraft as well. DARA has sought partnering arrangements with the industry, in particular with BAE SYSTEMS, with which it has now a joint venture on maintenance support for the Javelin VC10 air refuelling tanker. DARA provides support for the Jaguar, Hawk, Tornado, Harrier and Javelin fixed-wing aircraft (DARA 2003/2004, 12). Since 2000 DARA has a strategic partnership with Agusta Westland to support rotary-wing aircraft, for example for the Sea King helicopters. Agusta Westland (subcontractors BAE SYSTEMS and Thales UK) has been awarded a five year support contract for Sea King helicopters worth €438 million in a partnering agreement with the British MoD. Agusta Westland services the helicopters also during deployed operations. The Sea Kings are used for rapid reaction operations to transport Marines or for early warning. According to the NAO, there are presently five Sea Kings deployed in Iraq (UK NAO 2004). A similar arrangement of *Through Life Customer Support* has been made for the British fleet of CH-47 Chinook helicopters. Boeing recently won a contract worth £199 million over 34 years for supporting the helicopters (UK MoD 2006a).

Critics attest DARA no bright future, or any future at all. The MoD-hired financial advisors from Morgan Stanley recommend selling or closing down DARA altogether because it is not cost-efficient. According to Jane's Information Group, DARA was unable to secure new contracts, several older aircraft were prematurely withdrawn from service and the RAF has radically changed its way to support its modern jet aircraft (Jane's 2005). In the long run maintenance and logistics services will be provided completely as part of the supply-chain management of the original equipment manufacturers. The other option is to directly outsource particular maintenance services to the industry. For example, Northrop Grumman has been awarded a support contract for the E-3 Sentry AWACS aircraft by the British MoD. The company will provide maintenance services for a period of 21 years and the contract is worth $1.19 billion (Northrop Grumman 2004). A similar contract for maintaining Britain's fleet of C-130 Hercules went recently to Marshall Aerospace, which is worth £1.5 billion and runs over 24 years (Defense Daily 2006).

Maritime Maintenance

Traditionally private companies and shipyards have always played an important role in maintaining and overhauling military ships, and were also involved in logistics services. Also traditionally shipyards tended to build and service commercial, as well as military vessels. In fact, the origins of the American military-industrial complex can be traced back to the commercial shipyards, which built and maintained US Navy ships since the middle of the nineteenth century (Hackemer 2001). This shipbuilding industry, and in particular military shipbuilding, prospered especially before and during the Second World War and expanded throughout the Cold War. The end of the Cold War resulted in a steep decline of demand.

After more than 15 years of downsizing, there are now only six shipyards for servicing the US fleet and only two of them still build new ships. These are four US Navy shipyards (Portsmouth Naval Shipyard, Norfolk Naval Shipyard, Pearl Harbour Naval Shipyard, Puget Sound Naval Shipyard) and two private shipyards (Northrop Grumman/Newport News Shipyard and General Dynamics/Electric Boat). About 25 per cent of the maintenance and servicing work is done by the two private shipyards, which have an equal market share, and the remaining 75 per cent by the public shipyards.

The now Northrop Grumman-owned Newport News Shipyard (NNS) (acquired in 2001) in Portsmouth, Virginia, has built US Navy ships and done ship repairs and overhauls for the US Navy since the late nineteenth century. Since 1990 the shipyard has overhauled more than 200 US government and commercial ships. Northrop Grumman has the intention to strengthen NNS's military business and to provide more life-cycle services, as the orders for new ships have declined. The company is now an important US Navy repair and fleet supporter. NNS services the US Navy's nuclear carriers and submarines and is the only company capable of refuelling a nuclear powered aircraft carrier. It does the *Refuelling and Complex Overhaul* for all American nuclear carriers, which has to be done after 25 years of service and

NNS will provide these services to Nimitz class carriers over the next decade. NNS has numerous repair facilities on both the East and West coast of the United States. NNS's main private competitor is the General Dynamics subsidiary Electric Boat, which also maintains and overhauls US nuclear submarines, among them the new Seawolf and Virginia class boats. EG&G is responsible for the integrated logistics and engineering support of the Virginia-class boats.

In Britain the Royal Navy has outsourced a great proportion of its maintenance work. The Royal Navy has privatized its warship maintenance and support divisions in Faslane, Portsmouth and Devonport. Babcock Int. has a partnership agreement with the Royal Navy to manage the Naval Base Clyde (at Faslane and Coulport) for £400 million over five years beginning from 2002 (Babcock International 2003, 16). Fleet Support Limited, which is owned to 50 per cent by BAE SYSTEMS, provides maintenance and repair services to the Royal Navy at the Portsmouth Naval Base, which includes refits, upgrades and logistics. Devonport Management Ltd. (DML) owns and operates the Royal Dockyard in Plymouth, which it managed for the Royal Navy since 1987 and eventually purchased in 1997. DML is owned by Kellogg Brown & Root, Balfour Beatty and the Weir Group. The company is a Royal Navy prime contractor for life-cycle support to the Royal Navy submarines, which includes all support services from commissioning to disposal and also the refitting and refuelling of Britain's Trident nuclear submarines (N. Cohen 2003, 20). The original £576 million contract has in the meantime soared to over £900 million and the British government agreed to cover most of the cost-overruns (UK MoD 2002b, 3).

The Serco Group has since 2003 a partnering agreement with the Warship Support Agency, which operates 150 smaller vessels at Bases in Devonport, Portsmouth, and Clyde for which Serco provides maintenance and management services. In 1997 Serco Australia formed with P&O Maritime the Defence Maritime Services company, which has a ten year contract with the Australian Navy to service to service its fleet and to provide full life-cycle support for the new Armidale Patrol Boats (includes building) in an 18-year public private partnership. There is a whole range of smaller maritime companies in the UK, which offer maintenance and other technical services to the Royal Navy, but they mainly work as subcontractors to the aforementioned companies. Examples are British Maritime Technology, Flagship, Atkins Global and AMEC.

Depot Management

The US and Britain have already begun to privatize depots and depot workloads on military bases through various partnering agreements with the industry. The following section looks at the examples of the US, Britain, Germany and Canada, where depot management has already been partially privatized.

The United States Throughout the Cold War the US military had maintained a large maintenance, repair and manufacturing capability for almost all of its equipment, weapons and ammunition. When the Cold War ended and substantial force reductions loomed, there was the question how much of this in-house capability should be retained and how much could be privatized as a cost-cutting measure. Former Secretary of Defense John Deutch introduced the 'core capability' concept, according to which DoD would retain a maintenance capability at the depot level to support readiness and sustainment of all weapons systems needed in contingency operations (Clay-Mendez 1995, XI). Up to the mid 1990s the private share of maintenance and overhaul workloads even dropped. By 2001 half of the depots and arsenals had been closed down since 1987 (the peak in terms of personnel and workloads). In this year there were 19 depots of the formerly 38 left with many more base closures to be expected. The remaining depots have been downsized and are now working collaboratively with industry. The workforce in government-owned depots fell 59 per cent, which largely affects internal maintenance and repair capabilities. As a result the amount of contractor performed maintenance and repair has increased 90 per cent between 1987 and 2000. This means that the overall contractor share is approaching 50 per cent ($8.2 billion to $7.6 billion in 2000) (Warren 2001).

There are several US Army Commands in charge of managing depots and assigning maintenance/repair workloads: the US Army Material Command and subordinate to it are the Army Tank-automotive and Armaments Command (TACOM) and the US Army Field Support Command and Joint Munitions Command (AFSC/JMC), all of which use three different contracting arrangements for partnering with the industry. One possibility is teaming or work sharing. This is an arrangement in which one party is subcontractor to the other, which combines organic and commercial maintenance capabilities. Usually the depot is government-owned/government-operated, but part of the work is performed by contractors. Teaming is the most common arrangement in the US Army, as it allows depots to take advantage of cheap real estate, environmental permits, and cheap labour. For this purpose TACOM has set up the Ground Systems Industrial Enterprise (GSIE) in 2002, which is an umbrella organization that combines several Army depots in a partnering agreement with the industry (Irwin 2003) and which operates, according to its own mission statement, like a private business. This particular arrangement is used for supporting the Paladin artillery and Stryker armoured vehicles.

Secondly, private companies can buy goods and services from DoD depots and this includes also the resale of these goods to the DoD or other government agencies. This is an attempt to make public depots more cost-efficient and to use excess capacities for commercial work. And finally, there are leasing arrangements, where private companies lease public depots in order to perform maintenance and repair work there, either for the government or commercial business. The number of these 'government-owned/contractor-operated' facilities has increased. Examples are the Detroit Arsenal and the Lima Army Tank Plant operated by General Dynamics, the Lone Star Army Ammunitions Plant run by Day & Zimmerman Inc., the Holston Munitions Plant managed by long-term contractor BAE SYSTEMS and several other ammunitions arsenals, which are run by Alliant Techsystems.

The US Army has already done two major restructurings of the organization of its depots and arsenals since the 1980s and a third wave of restructuring lies ahead. The intention is to further simplify the organization and to split it up into only two components: field and national level maintenance. Additionally, more bases and depots will be closed. Former Secretary of Defense Rumsfeld proposed a list of 33 bases marked for closure, which represents about five to ten per cent of the US military bases (GovExec 2005). Base closures could be a way around the Congressional 50 per cent limit (Cahlink 2005). The Army would simply rely more the industrial supply chain and on just-in-time delivery for spares and maintenance.

Britain The MoD maintains depots for the repair of land vehicles and equipment, which are managed by the Army Base Repair Organization (ABRO). In order to cut costs and to increase the efficiency of ABRO, it has formed several strategic partnerships with industry and also made its repair facilities available for commercial business. The MoD spends about £290 million a year on repair and overhaul of land systems and part of the work is done in-house and part by the private sector. ABRO repairs 45 per cent of 15,000 repairable items and this represents about 60 per cent of the overall workload in terms of value, which leaves the private sector with 55 per cent of items and 40 per cent of the workload. The workload is partially allocated by competitive bids and partially non-competitively. An important reason for this is, as pointed out by NAO, intellectual property rights (UK NAO 2002, 5). Often the MoD does not own the intellectual property of its weapons design and the work has then either to be performed by the original equipment manufacturer or by ABRO, if an in-house repair capability exists. The MoD wants to increase the number of contracts awarded by competition to about 30 per cent, which is hoped to achieve higher cost savings. In principle, the MoD wants to retain an in-house repair capability of at least 35 key operational equipments and even those will be repaired in partnership with industry UK NAO 2002, 12). ABRO has entered strategic partnerships with Vickers, Alvis and BAE SYSTEMS. For example, under a new partnering agreement BAE SYSTEMS will support together with ABRO land systems like the future Panther command and liaison vehicle, the Challenger MBT and the Warrior IFV, which includes provision of spares and technical support UK MoD/ABRO 2005). NAO has therefore criticized the inconsistency of MoD policy in respect to the issue of retaining in-house repair capabilities for strategic reasons, which makes it unclear what levels of in-house capacity are actually required (UK NAO 2002, 28).

Germany and Canada The German Army has begun to privatize its maintenance depots in 2005. The newly founded Heeresinstandsetzungslogistik GmbH (HIL) [Army Maintenance Logistics Ltd.] is a joint venture between Krauss-Maffei Wegmann, Rheinmetall, and Industriewerke Saar, and will be responsible for the maintenance of some army equipment including tanks and armoured vehicles. HIL has been awarded a contract for a period of eight years, which is worth €1.77 billion. If this cooperation with industry is successful, then the German armed forces will consider further outsourcing initiatives. One of the official reasons given by the Bundeswehr for this decision to outsource important parts of its maintenance requirements is that it helps to sustain a healthy defence industrial base (German MoD 2005).

The Canadian Defence Forces have relatively little sophisticated material which would need intensive maintenance, but a report of the Auditor General in 2001 still complained that there were not enough maintainers to keep the equipment operational (Canadian Parliament 2001). In order to deal with these deficiencies, which affect the readiness of all three Canadian service branches, the Canadian Department of National Defence (DND) has now embarked on a policy of privatization. Under a new Optimized Weapons Systems Support programme that includes wheeled light vehicles (General Dynamics), the CP-140 aircraft (L-3 Electronics), and the CF-18 (Bombardier), the Defence Forces secured life-cycle support and management from private industry. Additionally, the Canadian DND has subjected its land forces depots to an Alternative Service Delivery (ASD) review, which will determine whether outsourcing would be more cost-efficient. Several armouries and facilities are already maintained by Atco Frontec, which was awarded a five year contract in 2003 for the maintenance of five additional armouries near London, Ontario (www.atcofrontec.com).

Focused Logistics

Modern armed forces are in nearly everything they do highly dependent on logistics services provided by private contractors. This section will give an overview on military privatization relating to transportation, base operations and weapons systems logistics management.

Transportation

Transporting troops and equipment into theatre and moving them around within the theatre has become one of the logistics functions that are now largely privatized. The problem is not new and the use of private providers of transportation for troops and equipment is not new either. Merchant ships have always been used in times of war to transport troops and supplies, D-Day being a famous twentieth century example. What might be new about it is the new relevance of the problem. As NATO forces were focused on the task of fighting a major war in Europe, they did not, with the exception of the US, need huge transport capacities for projecting force over huge distances. The 1991 Gulf War, in which no fewer than 735,000 coalition troops were assembled in Saudi Arabia and the Gulf region in just six months, was a historic logistical feat. 359,000 tons were airlifted and 6.5 million tons were sealifted into theatre, which are virtually mountains of equipment and supplies (Menarchik 1993, XIV, 2). However, it also made the logistical problems of deploying and supporting a large modern force quite obvious. In many respects the performance of military logistics was not satisfactory and needed to be largely augmented by commercial sealift and airlift. Without private companies providing the necessary sealift and airlift capacities, Desert Shield would simply not have been possible. Quite a few lessons were learned from that experience. Since then the US and other armed forces implemented new logistics methods and concepts taken from the private sector. They tried to improve military logistics with the help of private companies. Some of the most important logistics concepts that were copied from the private sector are: *just-in-time delivery* and *freight-identification.*

Additionally, because of the sometimes poor performance of military logistics capabilities, the American Transportation Command has contracted out the delivery of freight to commercial providers on several occasions when commercial companies performed better and were cheaper than military transport. In respect to the US operations in Afghanistan and Iraq following 9/11, commercial providers of transportation played a key role in the success of these operations.

The situation is slightly different for the other NATO members. Apart from Britain and France, no other nations have significant military sealift capacities and no other nation than the US has a sufficient military airlift capacity for quickly airlifting large numbers of troops for contingency operations. The following section looks at practices and examples of how Western armed forces augment their limited lift capacity by using private companies for providing strategic sealift, strategic airlift and intra-theatre lift capacity.

Strategic sealift Sealift is most important for getting supplies and heavy equipment into theatre. In the US there is the US Military Sealift Command (MSC), which manages a fleet of military cargo ships and military sealift operations. Additionally, MSC is augmented by commercial sealift capacity, which is ensured by the Voluntary Intermodal Sealift Agreement (VISA). Commercial cargoship companies can sign up for VISA and agree that they will provide a certain percentage of their overall sealift capacity for supporting contingency operations. Sealift in Desert Storm carried 95 per cent of all equipment and supplies and about 80 per cent were delivered by commercial ships (Menarchik 1993, 10). US Transportation Command claims that about 90 per cent of the US flag cargo fleet is part of VISA. The majority of the 95,000 tons of supplies and equipment sealifted during Operation Iraqi Freedom was provided by commercial ships under VISA.

Britain also has an arrangement with private industry to provide additional sealift when necessary. A large part of Britain's sealift requirements is met by the Royal Fleet Auxiliary, which is a government owned fleet of civilian manned ships. In contrast to the MSC ships, the British auxiliaries are armed and the civilian crews are under naval discipline should they be involved in warlike situations. During the build-up for OIF in 2002/2003, the MoD chartered two commercial ships, the *Sea Crusader* and *Sea Centurion*. Using a PFI the Royal Navy had already decided to outsource part of its strategic sealift to a private provider in 2000, but the ships were not yet ready at the time of OIF. AWSR Shipping Ltd. was chosen for the PFI contract to provide six roll-on/roll-off commercial merchant ships to the Royal Navy and operate them. The roll-on/roll-off vessels are needed for the quick global deployment of UK forces in contingency operations. The ships themselves have no military features, but they need to operate in war zones and they are crewed by Sponsored Reserves. During OIF the Royal Navy deployed four new roll-on/roll-off ships and chartered 60 additional commercial freight ships (UK MoD 2003b).

As Peter Singer has pointed out, the use of commercial shipping in support of military operations is not unproblematic, as illustrated by the case of a commercial ship company, which held a Canadian peacekeeping force hostage near Newfoundland in August 2000 (Singer 2003, 160). The peacekeepers were returning home from a mission in Bosnia and the shipping company refused to deliver the force to Canada because of a quarrel about the payment. Onboard the GTS Katie of Third Ocean Marine Navigation were five tanks, 580 army vehicles and 390 ammunition containers, amounting to ten per cent of the Canadian armed forces equipment, which was for a period of several weeks unavailable. The Canadian military had to eventually seize the ship by dropping soldiers from helicopters.

Strategic airlift The American Air Mobility Command (AMC) maintains a large fleet of heavy lift transport aircraft like the C-5, C-17, and C-141, which can provide lift capacity for up to one division. Despite this very impressive military airlift capacity, the AMC has to augment its airfleet with commercial aircraft during contingency operations, which form the US Civil Reserve Airfleet, an air transport equivalent to VISA. It was activated for the first time during the 1990 Desert Shield operation, where it made a major contribution for airlifting US personnel into theatre. An investigation by the US Congressional Budget Office states that 'During the Persian Gulf War, CRAF planes carried more than 400,000 personnel and 171,000 tons of cargo on more than 3,600 missions.' (US Congressional Budget Office 1997, 85) According to Douglas Menarchik, the relation of used airlift capacity was as follows: 20 per cent active duty, 30 per cent reserve, and 50 per cent commercial (Menarchik 1993, 11). The overall volume of charter contracts awarded in connection to Desert Shield/Desert Storm amounted to $1.9 billion. The report also acknowledges the role of commercial providers of airlift capacity in smaller contingencies. Commercial airlift played a major role in OEF and OIF. Very recently the US military has awarded contracts with commercial air freight companies such as FedEx to provide airlift capacity worth $2.25 billion, which is a significant increase to the previous year's (2004) airlift contracts worth $1.19 billion (Reuters 2005).

Other armed forces also make use of commercial providers of airlift capacity and charter aircraft on an on-demand basis. For example, part of the British peacekeeper force, which served in Kosovo, was airlifted by commercial charter planes. The British MoD was in the end rather dissatisfied with the use of commercial air assets, as it pointed out that 'The use of commercial airlift assets was constrained because it was unable to operate to all destinations, due to the risks involved and the difficulties in using aircraft from some countries opposed to the NATO operation.' (UK MoD 2000) Despite this criticism, it can be said that the limited availability of airlift capacity of most NATO members will leave them with few alternatives to hiring commercial aircraft when needed. This fact is acknowledged in Germany's initiative of creating a leasing/charter community for commercial Antonov-124 heavy-lift aircraft consisting of 15 NATO countries plus Finland, called Strategic Airlift Interim Solution (SALIS). Until the A-400M is available, the Ukrainian/Russian Antonov will carry the bulk of the military equipment of EU/NATO states into crisis areas (Koenig 2005).

Intra-theatre lift When troops arrive in theatre for a contingency operation in a Third World country their own transportation capacities are fairly limited. Often they have no other choice than to contract out transportation services to local companies, or to international companies, which specialize in moving military personnel and equipment around. For example, when the US forces deployed in Uzbekistan and Afghanistan, it was a major problem getting the supplies to them. The containers with equipment and supplies arrived in Karachi, Pakistan, and had to be distributed on trucks and transported over hundreds of kilometres inland. Additionally, commercially contracted 747s, C5s and C17s were used to bring more supplies into theatre. The bad condition of Afghan airbases often did not allow using big planes like 747s for airlift and some supplies had to be shipped from Germany through Russia all the way down to Afghanistan (Book 2002).

In a PFI Britain privatized the Royal Army's tank transporter unit in 2001. The FASTTRAX consortium led by KBR has agreed to provide 92 Heavy Equipment Transporters (HET), which will be maintained and operated by FASTTRAX in peacetime as well as during times of war for a period of 20 years for £290 million (UK MoD 2001). The MoD hopes to save 20 per cent of the costs through the PFI compared to normal procurement. HETs are needed to transport heavy tracked and wheeled vehicles to and from the battlefield in order to avoid wear and tear of the equipment and to save fuel. Heavy Main Battle Tanks like the M1 Abrams or the British Challenger 2 travelled most of the distances in the Iraq War on HETs. Therefore, the decision to outsource the HET unit means that a frontline service is privatized. The contractors, who operate the HETs, are Sponsored Reserves, however, which means that they have undergone military training and would become soldiers in times of conflict.

There is also a whole range of smaller charter companies specializing in military transportation. For example, ICI of Oregon provided intra-theatre airlift capacity in support of several international peacekeeping operations. The company mainly utilizes Russian-made transport helicopters and operates globally. In 1996 it helped to withdraw a Canadian peacekeeping contingent in Haiti and provided airlift capacity in Haiti for routine operations for the Canadian and Argentinean peacekeepers. In Nigeria, ICI provided aerial reconnaissance, medical evacuation and airlift capacity in support of the African Crisis Response Initiative to US and Nigerian forces. In a quite similar fashion, DynCorp has also catered for the peacekeeping market and provided four heavy lift transport helicopters for airlifting supplies for US and Australian forces in East Timor 1999/2000 (Collier 2001, 5).

Base Operations

Some domestic and international military bases were built and are largely operated by military contractors, which provide all the construction, engineering, management, security, and other technical services. This section will give an overview of private companies operating military bases in a domestic, international and contingency operations context.

The domestic context As a result of the shrinking size of the armed forces around the world, many military bases have been closed since the end of the Cold War. The military downsizing brought a whole wave of privatization initiatives in many NATO countries to further reduce the operating costs of the remaining military facilities and bases. Especially in the English-speaking world (in particular US, UK and Canada) many military bases are now largely maintained or supported by private contractors, who do everything ranging from mowing lawns to laundry services to maintenance on equipment and facilities. The US and the UK have also pioneered the privatization of military housing. The US Army has begun in 1996 to privatize military housing in several locations and by 2000 about 14 per cent of its housing stock had been privatized. In Britain a great proportion of the 547,400 married quarters were sold and in 1995 the Defence Housing Executive was created to manage the remaining 62,000 married quarters, including those which are now leased by the government (Pint and Hart 2001, XIV).

Base security is also a field where a lot of outsourcing is taking place. In Germany some military bases have been guarded by private security guards since the early 1990s and in Britain it is quite similar. In the US base security is already largely outsourced to private security companies like Wackenhut. According to the Los Angeles Times, 4,300 private security officers guard 50 Army installations in the US and the awarded contracts run as high as $1.24 billion (Miller 2004). In order to free more uniformed personnel for ongoing operations in Afghanistan and Iraq, the Pentagon plans to outsource base security for US military installations in Germany. This plan is also a result of the German decision to recall 2,500 German soldiers from guard duties of US facilities following the American announcement to withdraw 70,000 US troops from Germany (Hess 2005).

One US Army base within the United States, which is particularly dependent on contractors, is the National Training Center (NTC) at Fort Irwin, California, in the Mojave desert. Out of 8,000 military and civilian personnel based there 2,200 are contractors (Vinnell, Raytheon, and Johnson Controls Inc.). The NTC was originally built by Johnson Controls in the early 1980s, and it was decided that the company would provide all the base operations services for a 15 year period (Pint and Hart 2001, XVII-XVIII). Over 500 Johnson Controls contractors operate the base and maintain the vehicles and equipment. ITT got an additional logistics services contract in 1996 for five years for maintaining the fleet of 250 M1 tanks and M2 Bradley infantry fighting vehicles. ITT was recently beat out by Vinnell, which provides base operations services for $176 million per annum (Scholl 2002). Johnson Controls also manages base operations for the US Army in Fort Gordon, Fort Sam Houston, Fort Lee, and Fort Hood.

Britain's MoD has moved into the same direction and has recently awarded a £7 billion PFI to Aspire Defence Ltd. (a joint venture between KBR and Mowlem) to improve and service British garrisons at Aldershot and Salisbury Plain for a time frame of 30 years. This includes many important support services like catering and cleaning, as well as providing transport and reprographics.

Base operations abroad The need for troop deployments overseas has not diminished since the end of the Cold War and the US still maintains a large global network of military bases and facilities. Contracting out base operations for overseas bases is now rather the norm in respect to US forces and their global operations. As the possible examples are numerous, the focus will be on just two of the better known ones: *Camp Bondsteel* in the Balkans and *Camp Doha* in Kuwait.

Camp Bondsteel was built for housing 5,000 soldiers by KBR in 1999 to support US operations in Kosovo. KBR takes care of all the base maintenance work, provides, laundry, mail, food, and firefighting services, and also of supply operations and electricity generation amongst other things. The company did set up what some people believe to be luxury camp with enormous running costs and billed the US government accordingly. The construction costs were $350 million and the annual operating costs are estimated to be close to $50 million (Hackworth 2001).

Camp Doha was established as a US base following the liberation of Kuwait in 1991 and DynCorp International was contracted to maintain equipment based there. As the US wanted to keep large numbers of weapons and equipment prepositioned at the Iraqi border because of the possibility of a future contingency in Iraq, it had to maintain this large facility. Also coalition troops from Australia, Canada, Czech Republic, Britain, and many other nations were at times stationed there. In 1999 the international joint venture Combat Support Associates (CSA) was formed for maintaining the prepositioned equipment at Camp Doha and to keep the base operational until the contract runs out in 2009. CSA gets $51 million annually for base operations and maintaining tanks and other track vehicles in Camp Doha.

The US armed forces are not the only ones which use contractors for building and operating bases. This has, in fact, become quite common for international bases for training and contingency operations. Two examples are: the NATO training base at Goose Bay, Canada, which is operated by the British Serco, and Kabul International Airport, which is operated by AtcoFrontec. The Canadian air base Goose Bay was built during the Second World War and was a transit point for Canadian and US air forces to Europe. In the Cold War it became a large NATO training base and Goose Bay had to be extended in the 1960s and 1970s. After the Cold War Canada embarked on a policy of privatization, which included its military assets, called Alternative Service Delivery (ASD). In the mid-1990s the Canadian DND decided to use the instrument of an ASD for Goose Bay to cut costs and to improve the quality of the service. As the NATO partners (UK, Germany, Italy, and Netherlands) did not utilize the base as much as previously, and as the base was running on less than 50 per cent of its capacity, it was considered by the Canadian government to close it down completely. Following 9/11 the climate has somewhat changed and DND wants to keep Goose Bay. As a result Serco won a contract re-bid in 2003, which now runs for 11 years with a volume of $40 million annually (Serco 2003). Goose Bay is administered by a DND General Manager, but operated by Serco, which provides technical and non-technical support services, among them the maintenance of vehicles and emergency systems.

Another example of a private company operating an international military air base is AtcoFrontec's management of Kabul Afghanistan International Airport. After the end of the war against the Taliban, a large international/NATO operation was launched to provide stability in Afghanistan for a new democratic Afghan government. NATO's Maintenance and Supply Agency in Luxemburg awarded AtcoFrontec a contract to manage the whole airport and to provide all the information and communications systems and services related to it (air traffic control, system administration), the facilities operation and management (electrical, water and sewage systems), the maintenance of vehicles and technical equipment (servicing, repairs) and also engineering services (runway and infrastructure repairs). The airport is presently used by 17,000 uniformed personnel from 26 nations. Atco Frontec also recently (2005) secured a contract to support multinational NATO bases in Bosnia.

In the future the US government wants bring back more troops from overseas deployments, while at the same time handing over military bases and facilities abroad to contractors, who would maintain them and the equipment stocks on the bases. In case they are needed, troops could be brought into theatre with relative ease. Most troops would be withdrawn from Europe (Germany), but also from the Asia-Pacific region (Korea, Japan), and Africa (Klaus 2004). Chalmers Johnson claims that 'Camp Doha has become the army's model ammunition depot, a prototype and paradigmatic example of a forward base for "prepositioning" the equipment, ammunition, and fuel needed for a brigade-sized armored task force.' (Johnson 2004, 147)

Weapons System Logistics

The RMA is to a large degree dependent on a Revolution in Military Logistics (RML), as all these high-tech smart weapons and capabilities are highly dependent on equally advanced logistics systems. Therefore, it has been rightly pointed out that the notion of logistics supporting the warfighter is wrong. Rather logistics creates the warfighter by enabling him to utilize operational capabilities (Gluck 2002). To create and sustain certain high-tech capabilities for high-intensity warfare requires a very sophisticated logistical effort. It is necessary to keep hundreds of thousands of different spare parts available and to have the capability to routinely service and repair weapons systems, even during combat operations far away from the home base.

Stephen Ferris and David Keithly claim that twenty-first century warfare would make a completely new approach to logistics necessary: 'The RML requires the military logistics system to focus its attention on managing information and distribution rather than inventory. This focus will, in turn, require sophisticated technology to provide the necessary asset visibility, point-of-service tracking, automated materials handling and intermodal information networking necessary for a twenty-first-century logistics system.' (Ferris and Keithly 2001) All the logistics methods and systems employed by commercial logistics providers like FedEx or UPS, need to be utilized for military operations and to some degree this has already happened. This includes radio-frequency tagging and GPS tracking of freight, total asset visibility (awareness of all assets, including their status and location), just-in-time delivery and automated warehousing for items, which need to be kept on stock. By delivering supplies and spares only when they are needed, in the quantity they are needed, it becomes unnecessary to keep large and expensive stockpiles. The whole RML idea can only work through an intensive cooperation with the industry because the industry has to manage logistics for the military by producing and delivering the supplies and spares at the right time in the right quantity. According to Ferris and Keithly, it is already foreseeable that 'The services will no longer sustain the force structures to requisition, supply, and transport parts and equipment in the battlespace. Increasingly, third-party logisticians will assume many of these tasks.' (Ferris and Keithly 2001)

Private companies will take over all the crucial logistics functions including the overall logistics management. They are already developing logistics management systems and procedures, and are advising the armed forces on how to improve their logistics operations. The general trend is that the original equipment manufacturers more and more provide the necessary logistics services needed for keeping their systems operational. Sometimes specialized logistics companies take over the management function as well and help planning and coordinating logistics operations.

Total asset management Science Applications International Corp. has developed a Radio Frequency Identification (RFID) system for real-time total asset visibility for the Fleet Industrial Supply Center, Norfolk. Each item is listed in a computer database and every tag broadcasts a unique code, which allows the real-time tracking of the item. The company claims that 150,000 tagged pieces flow through that port annually, which can be tracked at any time. Additionally, the company is responsible for returning damaged equipment home for repairs, and it also has employees in Afghanistan and Iraq, who maintain and upgrade these logistics systems. RFID tagging, which was used in the Iraq War, solved a major logistics problem that the US forces had experienced during the Gulf War. At the peak of Desert Shield, the logistics system broke down because so many competing orders were made that it was impossible to track them all and to figure out what had already been sent and what still needed to be sent (Cordesman 2005, 205-207).

The most complex military logistics system is the Combat Support/Combat Service Support Army (CS/CSS-Army), which is based on a SAP application. More than 135,000 military and contractor personnel can monitor and manage the global logistics system, which is largely automated. The goal is to streamline operations from factory to foxhole. Other armed forces like the British and German forces also use SAP based logistics systems, which allows them to do more with less. Anthony Cordesman has made a good point by observing that 'What it cannot do is compensate for systematic underinvestment in sustainability and logistics systems - a problem that affects almost all nations in NATO as well as many other US allies.' (Cordesman 2005, 205) GAO has recently confirmed in a study the complete inadequacy of logistics and distribution systems during the 2003 Iraq War. For example, there were never enough trucks to supply the advancing Marine Corps and 3rd Infantry Division with even the most basic supplies such as food (US GAO 2005, 10).

Automated weapons logistics systems The Joint Strike Fighter was developed with the logistics in mind to support it. The aircraft comes with its own *Autonomic Logistics Information System* (ALIS), which manages all the weapons system logistics and even offers a tool for mission planning. ALIS identifies all the logistics and equipment requirements, forecasts logistics needs, schedules maintenance, orders spares, manages the supply chain, checks the availability of spares and tracks them in the global information system, checks that the right spares are delivered in the right quantity at the right place and even enables users to plan missions in accordance to available logistics capabilities. The whole JSF support system forms a global decentralized logistics network that allows a high degree of flexibility and a high level of sustainability. Lockheed Martin additionally offers training, maintenance support, diagnostics and prognostics, and a help desk. All this is at least promised by Lockheed Martin and at the moment little can be said about how the system performs in practice, as the JSF has not entered service yet (probably in 2012). It is quite obvious that Lockheed Martin plays the key role in providing the capability of a modern and versatile strike aircraft, not simply in the sense that it has designed and manufactured it. Robin Laird explains the completely new concept of the JSF. 'The JSF is designed to operate as a provider of missions in an integrated battlespace. The shift is from providing a pure platform to becoming a full-spectrum provider – one that is customer-driven and able to provide air combat capabilities, ranging from future new aircraft to full-service field maintenance.' (Laird 2003, 5) The only thing that the armed forces effectively contribute to this capability is the pilot (and maybe the air bases). All the rest comes from Lockheed Martin (and other suppliers and subcontractors).

Re-supplying troops on the battlefield Getting supplies and spares to the soldiers on the battlefield is much more difficult in respect to ground operations, as the suppliers have to operate in a hostile environment and are exposed to enemy fire. The recent experience in Iraq shows that it is a very dangerous business for the mainly civilian contractors, who drive supply trucks, to travel in military convoys because of ambushes with gunfire, improvized explosive devices and RPGs. Continuous disruption of the supply line could make the US military effort in Iraq unsustainable (Lang 2006). As functional logistical chains on the battlefield are crucial for mission success and as the armed forces have only a very limited transportation capability, the only way to get rid of civilian contractors for moving supplies to and on the battlefield is to use autonomous resupply systems. The US forces have already tasked Lockheed Martin to develop the Multifunction Utility/Logistics Equipment Vehicle (MULE), which would be a robot vehicle that could follow troops on the battlefield, resupply them with ammunition and equipment and evacuate wounded soldiers. Another option for logistics on the battlefield would be to use UAVs to deliver supplies. These autonomous vehicles will be unarmed or only lightly armed and do not have to fulfil complex missions. However, the task of developing unmanned ground vehicles is still very difficult. For now the US Army might simply remote control these unmanned resupply vehicles until the technology is available for autonomous ground vehicles (Lopez 2004, 65).

Logistics Management Services

Private companies have assumed a very extensive role in providing logistics and management services to the armed forces. Although not many of these services are actually technical in nature, they are simply too important to be ignored completely in an overview of military privatization. This section gives a brief overview of some major logistics privatization programmes, which have been introduced since the 1990s in the US, UK and Canada.

US LOGCAP contract The US Army awarded in 1992 for the first time the Logistics Civilian Augmentation Program or LOGCAP contract to Brown and Root Services. The contract was held by the company for five years and then it went to DynCorp, which held it also five years before Brown and Root (KBR) became again the primary logistics provider for worldwide combat support in 2001 for a duration of ten years (Cahlink 2002). The purpose of LOGCAP is to make life easier for Army logisticians, who would otherwise have to deal with a great number of private companies themselves in order to provide all the logistics for a major contingency. LOGCAP means that one main logistics provider takes the responsibility for the whole job of housing, feeding, and otherwise supporting a great number of soldiers. The services include (but are not limited to): supply operations (food, water, oil, ammunitions and many other items); field services (sanitation, laundry, food service, mail etc.) and other services which includes engineering/ construction/maintenance of facilities and infrastructure, transportation and medical services (Ehab e.a. 1999). The contract is cost-plus which means that the Army pays for all costs plus an agreed percentage (up to ten per cent) of the costs as a profit for the logistics provider. The logistics managed through LOGCAP provider KBR was considered a success in the case of US operations in former Yugoslavia where KBR helped the US military to cut the required manpower by almost half and save the US government a substantial amount of money (Bianco and Forest, 48). The services included: building 200 dormitory-style barracks in less than 90 days; providing 600,000 gallons of water a day and generating enough electricity to sustain a city of 25,000 people; running a supply center with about 14,000 product lines; washing 1,200 bags of laundry and cooking and serving 18,000 meals a day; and operating 95 per cent of the Army's transportation, including rail lines and airfields (Ehab e.a. 1999). In the case of the present contingency in Iraq there are now doubts about the rationale of relying completely on only one main logistics provider as the costs have skyrocketed and the service quality is not up to the agreed standard. An audit showed that a large proportion of Halliburton's charges looked suspicious (Witte 2006, A01). For this reason the US Army has decided to discontinue its Iraq contract with Halliburton and to hire three companies instead of one as logistics and another one to monitor their performance.

UK CONLOG contract In 2003 the UK MoD decided to pursue a similar approach to contractor logistics support as the US and introduced the Contractor Logistics or CONLOG contract, which is an arrangement for contractor logistics support to the Permanent Joint Forces Headquarters (PJHQ) for worldwide military operations and exercises. The first CONLOG contract has been awarded to KBR after an open competition for a period of seven years and is worth about £12 million, but depending on the operations conducted by PJHQ it could reach £50 million in a few years (Halliburton 2004). KBR is the prime contractor for the CONLOG contract and has subcontracted part of it to Serco, Inchape Shipping Services and International SOS. The size and scope of the British CONLOG programme is significantly smaller compared to the American LOGCAP, but in principle it is the same thing, managed by the same company. KBR has already started supporting British deployments in Iraq and Afghanistan and the company informs that its work for MoD in Iraq relates to camp infrastructure, catering and interpreters. The MoD hopes to gain both better, as well as cheaper logistics support by entrusting it to a single provider and at the same time stresses that this is meant to be an augmentation to military logistics capabilities, not a replacement of it (UK MoD 2004c).

Canada CANCAP contract Canada introduced a similar scheme in 2002 called Canadian Logistics Augmentation Program or CANCAP, which replaces the Contractor Support Program, started in 2000. The contract was awarded to SNC-Lavelin/PAE Government Services and is worth C$200 million, which might increase to C$300 million in the next few years, for a five year period with a possible extension to ten years. The services include base operations, communications and transportation and until now CANCAP has supported Canadian deployments in Bosnia and Afghanistan. The DND points at the reduced budget and size of the Canadian Forces as well as a higher operational tempo, as a reason for the introduction of CANCAP. Up to now the DND seems to be happy with the contractor performance in the CANCAP and it stresses the newly gained flexibility by utilizing private logistics providers (Canadian DND 2004).

Operating Equipment and Systems

Military operations require a great variety of support facilities and services ranging from weapons and equipment maintenance over base management to the operation of high-tech weapons systems and sensors. Most of these services need to be provided in the area of operations and a great proportion of them have already been outsourced to private companies. This section looks in more detail at cases where contractors actually operated facilities and equipment in support of military operations and where they have acquired in principle the capability to do so in the future. This section is divided into Radars and UAVs, other TECHINT, tanker aircraft and satellite launch services.

Operating Sensors: Radars and UAVs

James Manker and Kent Williams have noted that 'contractors increasingly are being called on to operate systems' (Manker and Williams 2004, 4). Most of the military systems, which are effectively operated by private companies, are sensors like radars and UAVs. Contractors sometimes build, maintain, operate, and even own the facilities or the equipment, and only their capabilities are leased by governments. Sometimes the facilities or equipment is fully government owned and largely operated by it, but contractors still provide vital operational support services. The general trend seems to point at government leasing of capabilities instead of buying specific services, as increasingly formerly government operated facilities are turned over to contractors. This makes contracting a lot easier for the customers (governments, armed forces), who no longer need to worry about the details in contracts.

Radars and early warning systems The privatized operation of radars and even whole radar stations is actually not a recent development. In the case of the UK it goes back to the 1960s, and even includes such a strategic capability as an early warning system for a Soviet attack. Being part of the US Ballistic Missile Warning System set up in 1959, the UK established an early warning radar (UK Air Surveillance and Control System) at RAF Flyingdales, North Yorkshire. The facility was then built by

Raytheon in 1962 and has been maintained and operated by RCA Services Ltd. (now Serco) since 1964. Serco has managed the facility for over 40 years (www.serco.com). This was actually the first outsourcing contract the MoD awarded to a private company. Britain had even outsourced another aspect of its early warning system. The AA motoring organization was tasked to get to the Prime Minister (if he was travelling on the road at the time) and inform him about an imminent nuclear attack (DeGroot 2004, 225).

In the US Northrop Grumman maintains, upgrades and helps operating the Defense Support Program (DSP) satellites at the US Air Force's Space and Missile Center in Azusa, California. DSP consists of two geostationary satellites, which use infrared sensors to detect missile launches and to calculate the trajectory and impact points of the missiles. Northrop Grumman has been the prime contractor for DSP since 1970 and it has recently been awarded an extension contract for its technical support services. DSP has been a crucial component of the US early warning system and will continue to be so until it is replaced later in this decade by the more sophisticated Space-Based Infrared System High and Low developed by Lockheed Martin (Richelson 1999, 223-236; Wirbel 2002, 67).

The Canadian Atco Frontec operates many military radars in North America. In 1998 the company won a $400 million contract to build and operate the Solid State Phased Array Radar System (SSPARS) for the US Space Command. The company also operates a large number of radar sites for the US government. Most radar stations in the US, Canada and Greenland which feed into the North American Aerospace Defense Command (NORAD) are now contractor-operated – a development that began already in the late 1980s. From five geographically separated locations the contractor-operated NORAD radar stations provide early warning against ballistic missile attacks, including attack assessment, to the US Air Force Space Command. Additionally, AtcoFrontec's SSPARS can track satellites and debris orbiting earth. In Alaska AtcoFrontec operates in a joint venture (called ARCTEC) together with Arctic Slope World Services Inc. NORAD radars at 18 different sites of which three are unattended short range radars. ARCTEC also operates the Northern Warning System (previously known as DEW Line) for the Canadian Department of National Defence where it provides, according to its website, 'airspace surveillance, attack warning, intercept control and navigational assistance to military and civilian aircraft'. AtcoFrontec has also operated mobile radar stations in support of military contingency operations, most importantly in the Balkans for NATO SFOR between 2000 and 2003.

Surveillance aircraft Contractors not only operate ground-based but also airborne systems like the JSTARS ground radar system and the Guardrail surveillance aircraft. In these cases the surveillance aircraft are owned and controlled by the government/armed forces, but in a few cases companies were hired to provide independent airborne surveillance capabilities.

JSTARS is an airborne ground radar system, which is packed into an E-8C (a modified Boeing 707) and monitors ground movements for targeting and battle-management with a field of view of over 40,000 sq km (Ebutt 2004, 257). The aircraft has a standard crew of 21 with four pilots/navigators and 17 crew members operating the JSTARS workstations. This system was introduced into the US forces just before the beginning of the 1991 Gulf War. During Operation Desert Storm JSTARS provided real-time battlefield intelligence on Iraqi mobile targets like armour formations and mobile SCUD launchers. The overall performance of JSTARS in the Gulf War is considered to have been very good. However, it was also acknowledged that the system is very complicated and demands a high level of skills from the operators, as it still has no automatic target recognition and identification capability (Ebutt 2004, 259). This means that operators have to identify targets cognitively. As the US military was not sufficiently trained on JSTARS, contractors had to sit side by side US Air Force personnel during combat operations in the Gulf region (Manker 2004, 14). Even during the last Iraq War, contractors were routinely flying on JSTARS aircraft (Singer 2004).

In the case of the Guardrail surveillance aircraft the US military has no in-house capability to maintain and operate it, as it is believed that it would not be cost-effective to do so considering the fact that the US Army has only a few of these aircraft (US GAO 2003, 9). Guardrail was even designed from the beginning to be completely contractor-operated (Guma 2004). The Guardrail system is a signals intelligence collection aircraft and is essentially a modified Beechcraft stuffed with antennas and electronic equipment, which can collect real-time SIGINT and targeting information. According to FAS, 'It collects selected low, mid, and high band radio signals, identifies/classifies them, determines locations of their sources, and provides near-real-time reporting to tactical commanders' (www.fas.org). Guardrail has been employed by the US Army for 25 years (with various upgrades and modifications) and recently supported OEF and OIF. This system (as well as the Navy's EP-3E) is planned to be replaced by the Aerial Common Sensor (ACS) by 2009, which will be fully interoperable with other military information networks.[1]

A very different case is Airscan. The Florida-based company mainly provides airborne surveillance services to other companies and governments. Founded in 1989 to conduct airborne surveillance for US Air Force launch facilities, the company has since then operated in Africa, the Balkans and the Middle East. Airscan operated a Cesna 337 Skymaster (the same type was used by the US Air Force in Vietnam) for airborne surveillance of oil fields in Angola and in Macedonia in support of NATO peacekeeping (Silverstein 2000, 176; Singer 2003, 163). Recently Airscan was awarded a $10 million contract for helicopter surveillance of the pipelines in Iraq. More controversial was Airscan's role in Angola and Colombia, where it indirectly interfered in ongoing conflicts by its intelligence collection function. The reason why Airscan is occasionally hired by Western armed forces is not so much the company's technical expertise, but simply because it can augment the usually scarce military intelligence capabilities.

1 Private companies will also play a key role in the ACS programme. The platform used will be a modified civilian regional jet equipped with advanced sensors for imagery and SIGINT collection and upgraded communications systems. Until recently Lockheed Martin held the contract for developing ACS. The contract was cancelled on the grounds that the programme would have exceeded the planned cost projections. But US Army and Navy still want to carry on with ACS, maybe with another contractor (Donnelly 2004).

UAVs Remotely piloted aircraft (or UAVs) have been used since the 1960s. They were usually contractor maintained and sometimes also contractor operated because they were quite unusual and very sophisticated pieces of military equipment. The first UAVs like the Vietnam era Firebee could usually only be used once and their film capsules needed to be recovered. With the development of real-time datalinks, UAVs now have assumed an increasingly prominent role in information age warfighting. The three most important ones presently in use by the US forces are the Hunter, Predator and Global Hawk UAVs. All three systems are relatively new.

Hunter entered service in 1996 and has since then been operationally used in Macedonia/Kosovo and Iraq. According to Linda Robinson, the TRW contractors did 70 per cent of the maintenance on the Hunter UAV during Balkans operations because the soldiers rotated in and out every couple of months and therefore never became proficient on the system (Robinson 2002). Hunter is presently undergoing testing to give it an offensive capability by pairing it with Northrop Grumman's Brilliant Anti-tank Munition (BAT). The General Atomix Predator B (Hunter/Killer) drone has already been used as weapons system in Yemen, Afghanistan and Iraq. The unarmed Predator A has been operational since 1995 and over 100 Predators have already been delivered to the US Air Force. During OIF contractors supported the Predators and their data links (Isenberg 2004), one reason being that the Air Force was not trained to maintain the entire vehicle (US GAO 2003, 2). Global Hawk, the biggest UAV in the American UAV fleet, had just finished flight testing when it was fielded in support of OEF in Afghanistan and later in the Iraq War. It flew over 50 missions in Afghanistan, a hundred in Iraq, with a total of more than 2,000 combat hours. Contractors from Northrop Grumman operated the Global Hawk UAV in Afghanistan (Petersen 2002) and Iraq (Singer 2004), which provided vital real-time battlefield intelligence in both campaigns. In the Iraq War Global Hawk operated out of the United Arab Emirates and flew combat missions every day of the war. It is unlikely, however, that the Northrop Grumman operators faced any additional risk, as the drone was controlled from a ground station at Beale AFB, California (Cordesman 2004, 307). The system itself is largely autonomous: it can take off, fly a preprogrammed mission and land all by itself. The technicians are mainly needed for troubleshooting and maintenance.

Other Technical Intelligence

Apart from operating and supporting groundbased and airborne sensors like radars and UAVs, private companies now contribute in many ways to technical intelligence collection. In fact private contractors no longer only provide technical services to the national intelligence organizations, but also have intruded other fields of intelligence work, which goes so far that the CIA has begun to even outsource its clandestine services. Former CIA operative Robert Baer[2] claims that since 1997 'practically all training is done by contractors' and that the 'CIA is even hiring contractors as station chiefs in other countries' (Chatterjee 2005). In respect to technical services, it can be said that there has been a very close cooperation between the intelligence services and the industry since the early days of the Cold War.[3] What is essentially new about the present situation, is that the intelligence services have lost their technological lead and rely now much more on the services of their contractors for developing, maintaining and operating their most sophisticated technical systems.

SIGINT With ever more and sophisticated technology, the intelligence services find it very difficult to stay ahead of commercial technology, as they are now facing private competition in fields which used to be the sole domain of intelligence services. The most important example is cryptology. Once almost exclusively used by governments and their intelligence services, this technology has proliferated to the business world, which has become increasingly concerned about the security of its communications. The development of advanced encryption methods in conjunction with the spread of the Internet allowed the introduction of E-Commerce and electronic banking and trading in the 1990s.

The side effect is that technology has become the enemy of the high-tech spooks. Michael Ignatieff claims: 'The once-powerful American intelligence community is now unable to compete with private companies in the development of new encryption software; nor can it recruit the most talented programmers into military service.' (Ignatieff 2000, 190) In the past the NSA could boast to be five years ahead of the commercial industry, but now it has almost lost the race with the private sector. According to Seymour Hersh, the NSA was largely unprepared for today's digital world with its high volume of data traffic and a significant proportion of it is protected by virtually unbreakable digital encryption leaving NSA's codebreakers far behind (Hersh 1999). The consequence is that many services relating to the collection and analysis of technical intelligence has been outsourced to private contractors. Because of the highly secretive nature of their work it is very difficult to determine which responsibilities these companies have taken over and what exactly they are doing. However, it is known that Raytheon not only develops the computer systems for the NSA, it also does top secret intelligence processing work at Buckley AFB, Colorado (Wirbel 2004, 102). Since the 1998 contract (revealingly named *Breakthrough*) CSC maintains and operates NSA computers for the agency's day-to-day operations (Guma 2004). Ironically, following budget cuts and personnel reduction, about 4,000 NSA employees switched to the private sector and ended up working for companies like CSC and naturally came as highly paid consultants back in. Another long-term intelligence contractor, SAIC, got several contracts

2 Robert Baer is now a well-known 'rogue' former CIA officer, who criticized the CIA harshly in his book *See No Evil* for its over-reliance on technical intelligence, its risk-aversiveness and its failure to respond to the threat posed by Islamic fundamentalism.

3 The NSA played an important, but unheralded role in the development of modern computers. In a huge government computer research programme it contracted (among others) IBM, General Electric, and MIT in 1957, 'extending the art of computer science beyond any expectations' (Bamford 1983, 136).

from the NSA worth hundreds of million dollars to revamp its eavesdropping systems by overseeing the Trailblazer project, but the project turned out to be a complete failure (S. Harris 2006). Nevertheless, SAIC also developed a datamining software (TeraText and Latent Semantec Indexing), which can monitor any kind of electronically transmitted communication (Chatterjee 2004, 160). Other important American TECHINT service providers to the US intelligence community are Anteon, CACI and Titan.

Satellite imagery At the end of the 1950s the US initiated a top secret space reconnaissance program with the name Corona Project. For many years the US and the Soviet Union (and later China) had a monopoly on space imagery. This was broken only in the mid-1980s when the French developed their own spy satellites. The main purpose of the Corona spysatellites was collecting strategic intelligence of 'denied' territories and since the 1970s the emphasis shifted to arms control verification. However, satellite imagery can also be of great importance for conventional military operations. A quite illustrative example is the case of the 1973 Yom Kippur War, where the US had provided the Israeli forces with satellite imagery showing the exact positions of the Egyptian forces. This enabled the Israeli Defence Forces to counterattack successfully (Ripp and Fontanella 1991, 15-89). Because of the great reluctance of the US prior to the 1973 war to supply valuable satellite imagery, Israel and some other nations became interested in developing their own imagery satellite capability. With the commercialization of space during the 1980s through Arianespace and the French SPOT Imagery company satellite imagery became commercially available.

The first SPOT satellite was launched with an Ariane booster in 1986 and SPOT 2, SPOT 3 and SPOT 4 were launched in 1990, 1993 and 1998 (Steinberg 1998). The French SPOT Imagery company offers commercially satellite imagery to anybody who can pay for it. Satellite imagery can be used for a wide range of applications from agriculture, environmental monitoring, archaeology, strategic intelligence and the planning of military operations. Of course, the quality of commercial imagery is far inferior to the capability of American and Soviet/Russian satellite imagery, but a high resolution is not necessarily needed or an advantage for military applications. However, certain imagery types like IR and radar imagery are only to a limited extent commercially available.[4]

There are certainly possible military applications for SPOT imagery, which has a resolution of 10m. Sandia National Labs claims in a feasibility study that it is possible to use SPOT imagery for the detection of massed troops in desert terrain, which exists in Africa, the Middle East and Asia (Gupta 2000). According to Gerald Steinberg, there are indications that both Iran and Iraq used SPOT imagery in support of their operations during the first Gulf War (Steinberg 1995). It is believed that Iraq used satellite imagery supplied by the French SPOT for planning and conducting the invasion of Kuwait in early August 1990. The US demanded that no further commercial satellite imagery shall be passed on to Iraq and the French government

4 IR and radar imagery is used to look through cloud cover and at night. IR imagery can also be used to find camouflaged targets.

indulged (Beason 1997, 61). In fact the American government itself purchased SPOT imagery for several million dollars during the war (Lewis 2002, 17; Seebode Waldrop 2004, 157).

Today there are now several sources of commercial satellite imagery apart from the French SPOT, most importantly the American Space Imaging, Digital Globe/EarthWatch (US), Orbital Imaging (US) and Israeli Aircraft Industries. Table 5.2 gives a brief overview of the most important commercial imagery systems and companies.

Table 5.2 Commercial and Civilian Earth Observation Satellites

Country	Company	Satellite System	Launch Date	Imagery Type
US	Space Imaging	IKONOS	1999	Panchromatic Multispectral
US	Digital Globe	Quickbird	2001	Panchromatic Multispectral
US	Orbimage	OrbView-2 Orbview-3	1997 2003	Panchromatic Multispectral
Israeli	ImageSat International	EROS A(A1) EROS B(B1-6)	2000	Panchromatic
Canadian	Radarsat	Radarsat 1 Radarsat 2	1995 2006	SAR(C-Band)
French	SPOT Imagery	SPOT 4 SPOT 5	1998 2002	Panchromatic Multispectral
European	European Space Agency	Envisat	2002	SAR (C-Band)

Source: Company Websites

As the market for commercial satellite imagery grew tremendously in the 1990s, the US government decided to regulate the operations of US commercial remote sensing satellite firms ('shutter control'). In crisis situations and in war time, the US government has the right to shut down the operations of these firms, if vital US interests are at stake. This has not yet happened and instead of restricting these firms, the US government seems to simply want to have exclusive access to sensitive commercial imagery during times of ongoing US combat operations. For example, the National Imaging and Mapping Agency (NIMA: now NGA) made a contract with Space Imaging (IKONOS), which granted it exclusive access to imagery on Afghanistan during the seven months of war in 2002. This is of course also a form of censorship (Baker, O'Connell, Venzor 2002, 2081).

Homeland security and domestic surveillance The US spends no less than $100 billion a year on homeland security ($36 billion is budgeted to the Department of Homeland Security (DHS)). The Homeland Security Research Corporation estimates the growth in spending from $5 billion in 2000 to $130 billion in 2010, which makes homeland security the fastest growing business segment in the whole US national security sector (Ismail 2004). Several private companies have already been awarded lucrative contracts for developing and providing high-tech systems, technical consulting and IT services. The US government believes that combating terrorism requires first and foremost high-technology, ranging from biometrical identification to systems for CB weapons detection and protection. In order to develop and utilize these high-tech counter-terrorism systems an Office of Total Information Awareness was created in 2002, which manages research projects for counter-terrorism and domestic surveillance. It relies heavily on at least five major contractors, most importantly Booz Allen Hamilton, Lockheed Martin, Syntek Technologies and CACI.

The biggest player in the homeland security field is probably CSC, which has provided computer software, systems and services to the US government since 1963 and which held government contracts over the last decade worth $15 billion. In 2002 the company got the $10 billion US-VISIT contract to provide systems and services for the Immigration and Naturalization Service (INS). CSC now supports INS to implement biometric scanning and identification, including fingerprint and facial recognition to identify terrorists and other criminals entering the US (Guma 2004). SAIC manages the computer systems of the INS for $1.2 billion (Harris 2006). Anteon International Inc. is the prime contractor for developing identification cards and biometric verification systems for the DHS's Resident Card and the State Department's Border Crossing Card programmes. The identification cards business is believed to be a fast growing business area. Also the British Home Office has very recently announced that it is going ahead with the introduction of the National Identity Card scheme. The Home Office wants to tackle a whole range of issues with it: terrorism, crime, illegal immigration, benefit fraud and convenient identification for every day transactions (UK Home Office 2004). The likely contractor for the Identity Card is the American Electronic Data Systems (EDS), which has a rather mixed track record in its dealings with the British government. As is the case with all IT projects, the job will not be finished by simply handing the equipment, but will require the contractor to do all the maintenance and troubleshooting and maybe some operational responsibilities as well.

BAE SYSTEMS currently tries getting a sizable chunk of the American and British homeland security business and works together with the DHS on solutions for tackling the threat posed to commercial aircraft from shoulder-mounted SAMs (MANPADS) (BAE SYSTEMS 2003, 17). Private companies also help to address the threat posed by NBC terrorism by developing detection devices, which can identify biological and chemical agents, by developing and manufacturing vaccines and by supporting first responders to an attack. For example, Battelle is quite strong in the NBC defence segment. Among other things, the company does defensive CB weapons research, advises the US government on domestic NBC protection and trains first responders (civilian and National Guard). Other countries are also tightening their domestic security and are introducing plans and systems for emergency management, which is a growing market for all kinds of IT and other technology companies.

Tanker Aircraft

The leasing of capabilities could reduce the initial investment costs for governments and would allow more flexibility for the armed forces. One major candidate for such kind of arrangement, which has been discussed since the late 1990s, are tanker aircraft. Boeing has pushed for a leasing deal with the US Air Force for its new 767 tanker aircraft with the arguments that the Pentagon has not the cash to buy the necessary number of tanker aircraft and that aircraft leasing is a quite common practice in commercial aviation. Critics observed that leasing the suggested number of 100 KC-767 would be significantly more expensive than buying them. The purchase would cost $8 billion (not including maintenance costs), while the leasing deal would amount to $15.5 billion over six years (Crock 2003). The whole thing looked like a rip-off. Boeing's main argument for the deal was that it would keep the 767 production line open and thousands of workers in Everett employed. Congress eventually cancelled the leasing deal in 2002, following the controversy over bailing out Boeing at the taxpayers' expense. The US Air Force also began an investigation of the matter in order to determine whether Boeing had violated conflict-of-interest laws (Pope 2003).

The Blair government has discussed privatizing the RAF's air-to-tanker refuelling fleet by using a public-private partnership. Despite the similarities with the failed Boeing deal in the US and the controversy that surrounded it, the British government plans for a contract with a private provider running 27 years for an estimated cost of £13 billion, making it potentially the biggest PFI to date. EADS is the preferred contractor, which will own, maintain and operate the tanker fleet and will provide the service during peacetime and the transition to war. If EADS gets the contract, the used platform will be 15 to 20 Airbus A330-200 planes with the first tankers are expected to enter service in 2010. The beauty of the proposed British arrangement is that the government has a viable exit strategy from the leasing deal. Should the aircraft not be used as air refuelling tankers, they can be converted to passenger planes or used as military transport aircraft. The partnering agreement would also allow to use any spare capacity for commercial purposes (commercial air freight) and to rent it out to third parties (Hartley 2004, 203).

Up to now, no definite decision has been made in the US or in the UK on the future of air-to-air refuelling capabilities. Boeing is still hoping to get some sort of arrangement with the US government for keeping its 767 tanker programme alive, while EADS tries to find some more customers and partners for its new Future Strategic Tanker Aircraft project. For example, the Australian Defence Force has already ordered five of the A330-200 and the Canadian and German forces have also recently ordered tanker aircraft. The leasing option for tanker aircraft outside Britain is not yet seriously considered by other Western air forces. In any case, it can be said that tanker aircraft are a strategic capability, which provide modern air forces with the necessary reach for contingency operations far from the home base. Many air forces either do not have sufficient air-to-air refuelling capabilities or need to replace their tanker fleets soon (the US tanker aircraft are in average 40 years old). The leasing of aircraft is quite normal in the commercial aviation and it seems likely that this option will become more widely used in the military aviation as well.

The British MoD already leases from Boeing (under the C-17 Leasing Company) four C-17 transport aircraft since 2001 for a seven year period, until the European A400M is ready for service. The British National Auditing Office (NAO) has praised the deal as an innovative and cost-efficient initiative of partnering with the industry (UK NAO 2002a, 30). The leasing of tanker aircraft and similar military capabilities could become a model for the future.

Satellite Launch Services

The space industry used to be even more than the rest of the defence industry under tight government control, as it was developing and employing the most advanced technology. Now NASA is one of the most outsourced and privatized government agencies in the United States. This significant change began already before the end of the Cold War with some first cautious steps, which were taken after the Challenger catastrophe in 1986. Slowly NASA increased the responsibilities of its suppliers and allowed them to take over small launch services in 1989. A contract for small launch services went to Orbital Sciences in 1991 and McDonnell Douglas got a contract for medium launch services in 1990. Joan Bromberg notes that 'In 1985, it was NASA policy to control US transportation into space, but by late 1988, NASA was beginning to buy its space transportation from private industry.' (Bromberg 1999, 159)

After the end of the Cold War the space industry was systematically commercialized and the services provided by private companies became much more extensive. Instead of delivering launch vehicles to NASA or the Pentagon, the manufacturers of launch vehicles (Lockheed Martin and Convair) now orbit research- , commercial- , as well as military satellites themselves. In 1996 commercial satellite launches exceeded for the first time government launches (Freedman 1998, 51). The satellite business has grown tremendously over the last ten years and by far the biggest growth relates to satellite services. According to the think tank CSIS, the volume of satellite services increased from $25 billion in 1996 to $40 billion in 2000, while satellite manufacturing amounted to $16 billion and launch services to $8 billion for the same year (Lewis 2002, 16). Despite largely increased demand for launch services for commercial satellites as well as its market size and growth rate, the commercial space industry is in effect not very healthy. Loring Wirbel points out that the number of satellite launches per year had shrunk from an average of 90 to 60 in 2001 and claims that only 15 of these were true commercial satellite launches (Wirbel 2002). The main sponsor of the space industry remains the US military and the War on Terror is the most important factor in the industry's recent recovery (Galace 2005, 19).

It is no surprise that the market for satellite launch services is dominated by American providers. In 1998 Lockheed Martin and Boeing presented their Evolved Expendable Launch Vehicles (EELV) Atlas and Delta IV, which replaced the expensive Titan IV and Delta launch vehicles. Both companies operate their own launch facilities in Cape Canaveral and Vandenberg AFB, California. Boeing also offers sea-based launches for the commercial satellite business. NASA, the NRO and the US Air Force routinely buy launch services from Lockheed Martin and

Boeing. Lockheed Martin made $3.8 billion on satellite launch services alone in 2004 (Lockheed Martin 2004, 31) and Boeing $2.9 billion in the same year (Boeing 2004, 46). Lockheed Martin formed a joint venture with the Russian space organization Khrunichev State Research and Production Space Center in 1995 called International Launch Services (ILS), which can launch satellites with Atlas and Proton boosters from Cape Caneveral and the cosmodrome Baikonour, Kazakhstan. ILS provides satellite launch services to the American and Russian governments as well as to commercial customers. In 2005 Lockheed Martin and Boeing formed a joint venture with the name United Launch Alliance, which allows the two companies to combine production and launch operations, which could save them about $100- $150 million annually. Excluded from the deal are ILS and the Boeing subsidiary Sea Launch.

Europe wants to retain an independent launch capability and access to space in the long term through Arianespace. The company offers launch services from Kourou, French Guiana, with the Ariane 4 and Ariane 5 boosters. According Bruce Elbert 'Arianespace pioneered the concept of a commercial satellite launching company, a path followed in the US by Lockheed Martin and Boeing' (Elbert 2005, 36). The world's first commercial space transportation company was founded in 1980 and has since then launched numerous military and commercial satellites for international customers. In the military segment Arianespace has launched the French SPOT and Helios reconnaissance satellites, the Xtar, MSG 3 and Syracuse 3B communications satellites, and also the British Skynet 4 satellites. The Ariane launch vehicles themselves are now manufactured by EADS and Arianespace has also entered a joint venture with the Russian Federation to operate the Soyuz booster from French Guiana by the end of 2006 (Arianespace 2004). Since 2003 Arianespace has an ongoing cooperation with Boeing Launch Services and Mitsubishi Heavy Industries (Launch Services Alliance). As Britain wants to remain a world class military power, it has its own military space assets such as the Skynet communications satellites. Recently Britain has developed its own imagery micro-satellite with a 2.5m resolution called Topsat. However, Britain does not have its own satellite launch capability and has to contract out satellite launches. Topsat, for example, was launched from the Plesetsk cosmodrome by the Russian Space Forces in October 2005 (BNSC 2005).

The predominance of strategic alliances in the satellite launch sector shows that the market is already tight and also the growing importance of the military/institutional market. The degree of international cooperation in the space sector, both military and civilian, is also quite high and there are sometimes rather surprising arrangements. For example, Russia is in the unique situation that its biggest spaceport (Baikonour) is located in Kazakhstan and that the Russian space infrastructure is partially owned by the US companies Boeing and Lockheed Martin (France 2000). One observer even claims that 'the US is critically dependent on Russian launchers' and that both countries can hugely benefit from further cooperation in space activities (Savelyev 2004, 99). Particularly interesting about the present situation is that such a strategic capability as the access to space is now almost completely left to private companies and that there are globally only few venues for orbiting satellites. Considering the importance of an unrestricted and responsive access to space (in case important space assets fail or are destroyed by ASAT weapons) for the success of military operations of the most technologically advanced militaries, the great reliance of them on globally connected private companies is surprising.

Conclusion

This chapter has mainly covered support services which are essential for military operations and which need to be provided near or even on the battlefield. The private provision of some of these services such as aircraft maintenance goes back to the Korean War. The privatization of some other service types, such as operating sensors or launching military satellites, is a relatively recent development. Sometimes in-house capabilities for servicing or even operating some military systems do not even exist or are inadequate. One of the reasons for this is that during the 1970s and 1980s Western militaries modernized without giving too much thought about the support needs of complex systems and the enormous costs of sustaining them (Demchak and Rochlin 1991). The result was that military in-house support capabilities were insufficient from the start. When the defence budgets became tight after the end of the Cold War a greater responsibility of defence industry in respect to maintenance and logistics was almost inevitable. New procurement projects now usually include life-cycle support from the beginning to address this problem.

This development has not only led to a greater contractor presence on the battlefield, but also means that contractors get more directly involved in military operations. In some cases they are already operating important military systems such as UAVs, sensors, and satellites. There is a particular dependence of militaries on privately operated space launch capabilities and space assets. Remarkable is the growing internationalization of the aerospace industries and the dependence of the armed forces on private or even foreign companies. Sometimes important military capabilities such as transportation and earth observation are commercially available on the global market. The dependency of the militaries on globally operating and sometimes foreign private companies is so high that it already raises serious concerns. The long-term implications of this military privatization are difficult to predict.

Chapter 6

Implications

This chapter will summarize the findings of the three empirical chapters and look at the implications. Military privatization has three distinct dimensions, which are of course closely connected, but nevertheless need to be discussed separately in order to understand the different mechanisms and dynamics of it. Technological change and resulting military privatization have military, economic and political implications. By dividing the phenomenon of military privatization into these dimensions it becomes obvious that there are largely conflicting rationales and aims at work, which often clash with each other. Overall, it appears that military privatization creates problems, if not dangers, in all these dimensions and that they may well outweigh advantages gained through it.

Contractors on the Battlefield

From a purely military point of view, some of the most important issues connected to the use of contractors relate to command and control. At the moment, there are lots of still unsolved problems and unanswered questions in regard to how the military can integrate contractors best in the existing military command structures and manage them to its best advantage. For military operations it is clearly imperative that contractors can be relied upon under any circumstances which might arise. One main problem is that contractors are civilians and therefore not subject to the uniform code of justice. They are outside the military chain of command and cannot be ordered by a military commander to do anything, especially not anything outside the scope of their contracts. Only the contracting officer can direct them to perform activities within the scope of their contractual agreements. This arrangement is problematic in many ways. First of all, not all eventualities can be foreseen when the contract is written (Costillo 2000). There could be circumstances under which a military commander would like contractors to perform other duties, which are not listed in the contract. In this case the contract would have to be altered, which is not easy in the field. A contracting officer would have to renegotiate the terms of the contract with an authorized representative of the contracted company. There are certainly situations in which either a contracting officer or an authorized company representative is not available and the military commander and the involved contractors have to decide on their own. If the contractors comply, it might incur additional risks and costs, which is problematic in case neither the military, nor the company wants to accept responsibility.

Who is in Charge?

Sometimes it is unclear who is allowed to direct contractors and what kind of authority contractors have in certain situations. Because of the age, greater knowledge and close contact to higher-ranking officers of some contractors, ordinary soldiers or even junior officers assume that the contractors have some sort of special authority and follow their suggestions as if they were orders (Cha and Merle 2004, A01). This confuses command structures and responsibilities and should it go wrong, it would be difficult to judge who is to blame. Peter Singer points at the disunited organizational structure contractor presence causes and claims that Private Military Forms (PMFs) 'are integral to but not within, the military operation' and that 'there are no established standards or even operating procedures' (Singer 2004c). This undermines, according to Singer, the concept of jointness as the whole of the armed forces is now composed of disjointed parts, which can hardly function as a unity.

Vulnerabilities

Contractors create new vulnerabilities for the armed forces, which can impact upon their military effectiveness. Contractor personnel are not required to meet the physical and mental requirements of military personnel. Although most contractors have undergone military training at some point in their career, this might not be the case in the future, as the pool of former military personnel shrinks with the downsizing of the armed forces (Avant 2001, 14). Contractors tend to be older and physically less capable than soldiers. Soldiers are permanently trained and their readiness is ensured in various ways, but 'there is no system currently in place to monitor contractor readiness' (Orsini and Bublitz 1999). Their unchecked health condition poses an additional risk to the success of military operations. They tend to be less well equipped than the military in terms of protective clothing such as bullet proof vests or NBC protection and they are generally not allowed to carry weapons, especially heavy weapons. It appears that contractors have been preferred targets for terrorism in Iraq because they are soft targets (Manker and Williams 2004, 6).

More serious still is the issue of possible NBC attacks. While soldiers have to undergo regular drills in NBC protection and have gas masks ready at any time, contractors are usually not required to do this. It can be assumed that contractors would suffer most casualties in the event of an attack with NBC weapons. This was at least shown in a US wargame called *Global 95*, which simulated a North Korean attack on South Korea using chemical and biological warfare. The outcome of the wargame was that the US and South Korean forces were not able to stop the North Korean troops conventionally because they relied operationally far too much on civilian and contractor support which quickly fell apart after the CB warfare attacks and which drastically eroded operational readiness (Mangold and Goldberg 2000, 326). Equally important as the physical impact of WMD is the psychological impact. WMD can cause so much fear that this undermines the morale of military and civilian personnel. As civilians are less protected and contractors cannot be forced to stay in theatre when a NBC threat emerges, they might feel compelled by fear to leave with enormous consequences for the operational readiness of the military forces.[1] Some weapon systems require contractor support and if it becomes suddenly unavailable, it can degrade military capabilities as soldiers are left with equipment they do not know how to maintain or even to operate (Yeoman 2003, 38-47).

Another problem in respect to contractor reliability pointed out by Christopher Spearin is that contracted companies often hire indigenous personnel because it is cheaper and allows them to have more boots on the ground. This causes serious oversight problems as the Pentagon now no longer knows how many contractors it employs (Mandel 2006). Background checking of subcontracted personnel is often insufficient, with the result that it 'brings potentially hostile elements in the close with American forces' (Spearin 2003, 40). This means that the enemy can easily infiltrate the armed forces to gather intelligence or to carry out terrorist attacks. The coalition forces experience this problem in Iraq, where insufficient background checking has brought enemies into Iraq's new security forces. This not only hampered the effectiveness of these forces, but infiltration seems to be responsible for the success of some of the most violent attacks (SIPRI 2005, 116).

Contractors also reduce a military commander's flexibility, as he can only use them for clearly prescribed activities and has the obligation not to expose civilians to unnecessary risks (Campbell 2000). They are unarmed[2] and not supposed to take part in combat. If soldiers are used for logistics or other non-combat duties, they can still be used as additional fighters, if circumstances make it necessary (Singer 2003, 163). Instead contractors need to be protected, which requires additional military personnel that cannot be used for other duties. Having contractors accompanying military units into combat therefore causes additional burdens for the military, which have to be balanced against the advantages of having contractor support on the frontline (Orsini and Bublitz 1999).

A problem discussed by Peter Singer is that private companies tend to use less secure technology when they are contracted to temporarily fill military capability gaps. For example, the communications they use are generally less secure because of lower grade civilian versus higher grade military encryption (Brown 1999, 22). Sometimes their communications are not encrypted at all, which could compromise the success of military operations (Singer 2003, 163). There is also the issue of the military use of commercial items by the military and contracted companies. For example, GPS receivers are equipped with a limited availability function in order to deny the enemy the use of GPS. However, as the US military was short on GPS receivers during the 1991 Gulf War it made the decision to switch the limited availability function off, so that it could use commercial GPS, which also theoretically had enabled the Iraqi military to use commercial GPS as well (Correll and Nash

1 During the so-called 'tree-cutting' incident in Korea in 1976, when two US officers were killed in the demilitarized zone, US forces were put on DEFCON 3 alert. This resulted in hundreds of DoD civilians, who had replaced US soldiers for maintaining depots in Korea to request evacuation from South Korea in anticipation of a major war (Orsini and Bublitz 1999).

2 Some American systems contractors are allowed to carry a 9mm pistol for self-defence, but this depends on company policy and the decision of the military commander.

1991). Commercial electronics equipment used by contractors is also not hardened against EMP, allowing an enemy to shut down contractor-enabled capabilities with EMP weapons.

Military Ethics

The widespread contractor presence in the military has contributed to a process which could be called the 'civilianisation' of the armed forces. The whole process raises serious questions about the future of the military profession, which is unique because of its particularly high ethical standards (Snider and Watkins 2000, 5-20). With contractors intruding in the military realm 'the military's professional identity and monopoly on certain activities is being encroached on by the regular civilian marketplace' (Singer 2005, 119). It is not just the presence of civilian contractors that causes concerns, but the fact that they are already so much involved in military training and education. As military 'officers spend as much as a third of their careers in classrooms' (Cohen 2000, 53) there are serious worries that privatized military education could affect military professionalism (Avant 2005, 118). Contractors often work very closely together with soldiers and sometimes contractors are so much integrated in a military unit that they can feel as being part of it.

However, there are still great differences. While soldiers are supposed to follow military professional ethics, which basically means a willingness to put their own life on the line, if the mission or their commander demands it, contractors do not need to have such a high ethical standard. They are just employees and their main motivation is financial compensation for their work. For many soldiers it is demoralizing that some contractors do a similar job as they, but are much better paid, enjoy more freedom and can back off whenever they choose.[3] At a certain level of frustration a well trained NCO or officer is easily tempted by the lucrative offers of some defence and security companies. According to the head of US Special Operations Command General Bryan Brown, the 'Competition with the civilian world has never been greater' (Kennedy 2004). At the moment Special Forces members are the group with the best opportunities in the private sector. As a result, they are resigning from the armed forces in record numbers to take on the well-paid security jobs in Afghanistan and Iraq (AP 2004b). As the training of a special operator can easily cost a million dollars, this represents a serious financial loss for the armed forces (Schmitt and Shanker 2004, A1).

Private sector opportunities and large contractor presence could in the long term effectively undermine the high standards of military professional ethics and lead to a general mercenary attitude among those who serve. Peter Singer therefore calls the logic of military outsourcing 'self-defeating from the military's perspective' (Singer 2005, 119-133). Marketization forces the military to buy back the skills lost to the market from the market to a higher price with a potential loss in morale and combat effectiveness.

---

3 Michael Lanning points out that a contractor commonly earns four to five times the wages infantry privates or specialists earn and that a contracted truck driver often earns more than a junior officer (Lanning 2005, 229).

Loss of Competence

There are serious long-term consequences for the armed forces in relying too much on the private sector, especially when it comes to technology. Private providers of simple logistics tasks or even security services contractors could be easily replaced by any other contractors or by military personnel, if there was sufficient political will to do it. This is not so easy in respect to some technical services in support of military high-tech systems. The market for military high-tech is relatively small and the competence for deploying, operating and maintaining such systems is correspondingly scarce. There are only few companies that can do maintenance and repair work on an Apache helicopter, or an M1 Abrams tank, or the Aegis air defence system. If one contracted company suddenly cannot or no longer wants to fulfil its contract obligations, it might be very difficult to find a suitable replacement, especially if it is at short notice. In some cases only the OEM can perform the necessary tasks related to operating certain pieces of military equipment, as no other company has the competence.

The US GAO has therefore criticized in its report that in cases where the armed forces depended on contractor support there was often no back-up plan in case contractor support was not available (US GAO 2003). In its investigation GAO found that military commanders often simply assumed that they could replace contractors who have become unavailable with other contractor or military personnel, but there were no firm back-up plans (US GAO 2003, 3). The issue of contractors failing to perform according to contract obligations is not a purely academic one, as the experience of Desert Storm and the 2003 Iraq War has shown. Several contractors who were under obligation to deploy did not show up, leaving the deployed commanders with a degraded ability to perform their missions (Wood 2003).

The reliance on contractors through a loss of competence will undoubtedly grow in the future because the armed forces have already begun effectively to eliminate entire military career paths by transferring their responsibilities to the private sector. Most affected are probably logistics officers and maintainers. Lourdes Castillo has pointed out the long-term consequences of experimenting with outsourcing: 'If, after a five- or 10-year trial period, the concept does not prove successful, the military will find itself unable to instantly grow, train, and benefit from the experience of the mid- and upper-level managers now developed within the enlisted and officer corps. It will take close to an entire career of 20 years before the military can regain the capability now resident in its personnel.' (Castillo 2000) As the US and British forces have already awarded many long-term contracts extending to more than 20 years to the private sector, there seems to be little chance of reversing the general trend.

With the outsourcing of complex activities the military loses in the long term the competence not only to perform these activities itself, but also to monitor them effectively. This problem is exacerbated by the ever growing complexity of weapons systems. Procurement projects have already become so complex that nobody really understands them. The result is that the risk of failure of procurement projects has substantially increased while cost overruns have become routine (Kaldor 1983, 52). Very illustrative is the Army's Future Combat System (FCS) programme with projected costs in excess of $100 billion.[4] The US Army does not have the in-house engineering and systems integration expertise for managing such a complex project and handed the responsibility to the private sector, namely to Boeing. This has practically led to a near absence of oversight, which puts a big question mark on the success of FCS. The project is based so much on immature technology that the Army estimates the chance of success at 70 per cent. The fact that Boeing struck a deal, which allows it to make money even if it cannot deliver, does not help either (Royce 2004). There are serious dangers, if the buyer of goods and services ceases to be a 'smart buyer'. In the commercial sector the market can compensate to a certain degree for the lack of competence on the part of the buyer: the market pre-structures the options for the buyers in a process of market selection. In the defence business there are few market mechanisms that ensure quality and keep the prices low, as there is little transparency and competition. A significant portion of contracts has to be awarded non-competitively anyway for legal (patents) and competence reasons. The result is that the buyers do not understand what exactly they are buying and have afterwards few means to hold the contractor accountable for shortcomings. This raises not only costs, but also can cause capabilities to be unavailable, when needed.

It is important to understand the military as an array of various capabilities, which are not simply there, but need to be created for any specific task in which enabler and soldier have to form a unity. The 'weapons system', which represents certain capabilities, is not just a complex piece of metal that is operated by a single soldier, or small crew; it is just the sharp end of a large support network (Kaldor 1983, 7-20). Without the enablers the specific capabilities for the specific task do not exist, rendering the soldier, who depends on these capabilities for fulfilling a mission, unable to do so. The more weapons and units form a network, the more individual capabilities depend on the network as a whole and, in fact, need to be created by the network every time they are needed. To understand the immense role of private companies in that overall network it might be helpful to refer to an example from the civilian world. Companies sell mobile phones, but the buyers of mobile phones cannot do anything with them without the services connected to them. The overall capability of being able to make phone calls anywhere is created by the network. When the phone is used it becomes even physically part of this network. Without the network it is just a useless piece of plastics and electronics. Similar are the newest weapons like GPS guided bombs or UCAVs, which are pretty useless as single items, but very powerful as part of a network. As private companies are those who design and operate the network, they are the ones who have to create the military capabilities when they are needed, which hands over an immense responsibility to private organizations. To stick to the mobile phone example, you could say that the military is still making the calls, but the private companies are those which put them through. As any user of mobile phones will know, it leaves one at the mercy of the network provider in terms of service availability and tariffs.

---

4 The British have a similar, but a lot more modest programme called *Future Rapid Effect Systems*, which is meant to replace present light and heavy land systems with a range of more flexible, modular, and networked medium weight systems.

Legal Issues of Military Privatization

The question of the legal status of military service contractors, both under national and international law, is of great importance, as this prescribes and limits their role, rights and duties. Although the phenomenon of military contractors who support military operations is not new, their legal status is now insufficiently defined under international law, as they have taken on many functions, which are indeed new. This means that national, as well as international law needs to be revised (Zamparelli 1999, 24-26). Their use has also become more widespread and their role more extensive, which raises serious legal questions of what they should be allowed to do and to what extent they or the governments that employ them, should, or can be held accountable for their actions. The legal status of contractors deployed in a theatre of operations is not yet clear and widely discussed in the literature. Under international law there are four possible statuses for military contractors: they can be considered mercenaries, non-combatant civilians, lawful combatants, or illegal/unlawful combatants.

Can military service contractors be considered mercenaries? This question has arisen mainly because of the boom of private security companies (PSC) and private military companies (PMC) in the 1990s and is now quite relevant to the situation in Iraq, where such companies now are used to uphold security and to take on quasi military roles, like escorting convoys, guarding buildings, or training the new Iraqi security forces. Many former professional soldiers were lured by PSCs and PMCs to Iraq for very generous compensation that it seems self-evident that they are rightfully termed mercenaries, at least from a moral perspective. In respect to other military contractors, however, this is much more difficult to establish. In legal terms it can be said that mercenarism is internationally outlawed by a convention of the OAU and the 1989 UN 'Convention against the Recruitment, Use, Financing and Training of Mercenaries', which would mean that they would not be treated as lawful combatants, but rather as criminals. Additionally, many states have a national legislation against mercenaries, which clearly shows that they are generally not considered legitimate. In practice, however, this seems to be a different matter (Davies 2000; Adams 1999, 103-116). The criteria for mercenaries under international law are both too general and too narrow in order to apply the law, as there are plenty of loopholes to allow anybody to exclude oneself from that law.[5]

Are military contractors non-combatant civilians? Military contractors usually insist that they are non-combatant civilians and refer to the 4th Article of the Third Geneva Convention, which grants civilians accompanying force protection under international law.[6] It is not permitted to target them deliberately and if taken prisoner

---

5 In regard to the Geneva Conventions Article 47 David Shearer quotes Geoffrey Best, who said 'any mercenary who cannot exclude himself from this definition deserves to be shot – and his lawyer as well.' See Shearer 1998, 18). The 1989 Convention is now ratified by the necessary 22 states, but essentially remains a dead letter.

6 '... persons who accompany the armed forces without being members thereof, such as ... supply contractors, members of labor units or of services responsible for the welfare of the armed forces.'

they are entitled to be treated as Prisoners of War. They are, in fact, allowed to carry small weapons for self-defence purposes, but they are not supposed to participate directly in hostilities.[7] In the past this simply meant that contractors could be used in the rear, but were not supposed to work near the frontline. With the extended battlefield, non-linear operations and asymmetric warfare the distinction between a safe rear and a hot frontline has lost all its relevance. Not surprisingly, contractors can now be found increasingly on the battlefield. They sometimes provide crucial support for combat operations, which puts them at risk being considered by the enemy not only as legitimate targets, but also as 'unlawful combatants', who have no rights under international law.[8] This risk is heightened by the penchant of many contractors to wear uniform-like combat dress, which makes them virtually indistinguishable from real soldiers (Carter 2004).

Technological blurring Contractors are sometimes acting as *quasi-combatants*, for example in the case of security service contractors in Iraq. This raises the question whether they have to be considered lawful or unlawful combatants. The Fourth Hague Convention and the Third Geneva Convention have a loophole for non-state forces like guerrillas and militias, which could also grant contractors lawful combatant status, if they fulfil certain conditions. They need to 'be commanded by a person responsible for his subordinates', to 'have a fixed distinctive emblem recognizable at a distance', to 'carry arms openly', and to 'conduct their operations in accordance with the laws and customs of war' (Faite 2004, 171). If military contractors take part in hostilities and do not fulfil these conditions, then they would have to be considered 'unlawful combatants' and any nation which uses them in this function would violate international law. In the case of systems contractors the case seems to be clear: they cannot be 'lawful combatants', as they are outside the military command chain. Therefore, contractors are clearly not allowed to crew and operate a weapons system (Campbell 2000). A particular legal problem is caused by technology, as it increasingly blurs the distinctions between combat and combat support, intelligence and action, and combatants and non-combatants (Stanley-Mitchell 2001, 267). If contractors operate an UAV in support of combat operations, which provides real-time intelligence to military forces for locating targets and target acquisition, then they are directly taking part in hostilities. On the other hand, it is not even clear whether an UAV that acquires targets for shooters has to be considered a weapons system, or not.

Legal accountability of contractors The present privatization was not a very well planned and organized process, which in effect means that many functions have

7 Alexander Faite points out that just the possession of a weapon by a civilian in a war zone does not make him automatically a mercenary or unlawful combatant, as the weapon could serve a whole lot of other purposes unrelated to the military conflict, e.g. protection against crime (Faite 2004, 173).

8 Christopher Spearin points out: 'Had Iraqi forces captured some contracted personnel in the field, they might have been able to deny them legal rights either on the basis of what the contractors did specifically or perhaps through a blanket assumption informed by the changing nature of contractor activities.' (Spearin 2003, 38).

been too hastily outsourced without checking the accordance with national and international law. In order to address some of the issues, the American Congress has passed the Military Extraterritorial Jurisdiction Act (MEJA) in 2000, which basically extends the Uniform Code of Military Justice to contractors on deployed operations, so that contractors who commit crimes overseas can be brought to justice. Many critics of PMCs have pointed out that criminal and even heinous behaviour of military contractors is an issue.[9] Unfortunately, MEJA seems to be unsuitable for solving it because contract law stands against MEJA and there are conflicts with the US constitution as well (Campbell 2000). Peter Singer also pointed out that four years after MEJA was passed no contractor had been held accountable under it (Singer 2004). It is also not clear to what extent the armed forces have a legal obligation to provide security to their contractors and how much physical protection contractors can legally demand (Ferris and Keithly 2001. 9).

A possibility to solve the legal problem could be the British model of 'sponsored reserves', which allows military contractors to quickly switch from civilian status in peacetime to military status in wartime. The downside of this model is that it does not work for all contractors and all kinds of services. Some contractors do not like the idea of being reservists with all the obligations this implies and there are also international contractors, who are foreign nationals and who cannot be legally compelled to become reservists. It also only makes sense for services which are more military in nature and which demand a forward presence of the contractors. Other services which are more comparable to civilian work, but still need to be provided near or on the battlefield, or generally in a war zone, would not naturally be considered as candidates for the 'sponsored reserves' model, for example the setting up and servicing/maintenance of computers and networks. As a result, there is currently no overall solution to the status and responsibilities of contractors and their employing agencies.

The Economics of Military Privatization

The second chapter has already introduced the economic side of military privatization and has explained the logic behind the so-called Revolution in Business Affairs (RBA). This section will scrutinize the RBA argument and will look at the economic mechanisms and implications of the drive towards privatization. In a first step the understanding of 'war as business' will be questioned. Secondly, the arguments for and against cost-savings will be discussed.

War as Business

There is now a strong sentiment among politicians and some military leaders as well that the military should use its resources much more efficiently, especially as they are shrinking. Cost-efficiency is certainly not a bad idea for any organization with

9 There is the often-cited case of DynCorp employees who forced girls to prostitution in the Balkans, the DynCorp supervisor raping two of them. Apparently, none of them were prosecuted (Singer 2003, 222).

limited resources, but the argument goes as far as to suggest making the military operate according to business principles and almost like a business organization. Costs and tight budgets seem to be a main reason for this, but there are other reasons as well, for example the growing proximity between business and military worlds. The military depends on private companies for almost anything it does and it is quite natural that the way these private companies operate influences the way the military operates.

However, the pursuit of efficiency is highly problematic for the armed forces in times of war. War is quite the opposite to efficiency and productivity: it is generally wasteful and destructive. The objectives of war and business are indeed a far cry away from each other. An armed force needs to be effective, not efficient, if it wants to win. Of course, it is a good idea for the military to manage its resources carefully and to apply sufficient, rather than always overwhelming force in order to avoid waste. However, Clausewitz has pointed out that there cannot be any limits to the use of force in war, as the enemy can escalate the resistance. War is not similar to business competition because of this possibility of endless escalation. There are of course rules in warfare, which limit the horror and destructiveness of war, but it is also true that there are no umpires in war and every side can break agreed rules of war, as long as it wins. This makes war potentially unregulated by rules and also chaotic and unpredictable. While in a market the behaviour of the agents within this market follow agreed rules, which makes the concept of a market possible in the first place, this is not the case in war. Rule-based behaviour has also the effect that markets themselves follow rules to a certain extent. Sure, markets are not very predictable either, but they tend to be quite stable, while war has a strong tendency to the extremes of victory or defeat.

Additionally, war cannot be itself profitable – at least not in our times. It has been often pointed out that the cost of war has substantially increased, especially in the twentieth century, while the possible economic/political gains were diminished (Mandelbaum 1998, 20-38). If it was true that we would be going back to the Middle Ages with war entrepreneurs as the main actors in warfare (as claimed by van Creveld 1991; Münkler 2005; and to some extent by Kaldor 1999) it would be necessary to make warfare a profitable economic activity and there are only three main possibilities for this: 1) plundering wealth from the population; 2) slave labour; and 3) exploitation of natural resources. From these three only one could yield any chance of making a profit out of war for a business organization operating in an international legal framework and that is the exploitation of natural resources. Even this is highly questionable, as more recent examples illustrate. The PMCs Executive Outcomes, Sandline and Levdan have all tried to get a stake in the countries' natural resources for which they worked, but usually failed to make large profits (Chatterjee 1997, 27-37). Madeleine Drohan showed in her analyses of cases where private companies used force to help their business interests that their efforts rarely paid off (Drohan 2004).[10] Although some multinational companies operating in Third World countries sometimes maintain 'private armies' to protect their property and

10 I am mainly referring to the examples of Union Miniere in Congo, Shell in Nigeria, and Talisman in Sudan.

personnel, security can quickly become such a substantial cost factor that it forces these companies to withdraw from a country.[11]

The costs of war have always been very high and only where the means for waging it are sufficiently cheap can private actors become a party in war.[12] The reason for this is that defence is both very expensive and a public good and therefore usually needs to be financed through taxation (Kennedy 1983, 23-31). It is the nature of public goods that they do not allow exclusion and there is no rivalry of consumption. Once a public good is provided, individuals cannot easily be excluded from it and the good itself is not diminished through consumption. Private business, however, works on the principle of exclusion. Those who do not pay for a commodity or service can be excluded from it. With public goods there comes naturally the problem of the 'free rider': there are little incentives for individuals to contribute to the costs of a public good, as the benefits of it can be enjoyed anyway. Governments can deal with the free rider problem by financing public goods through taxation. Therefore, private companies cannot provide public goods, as they cannot demand taxes like governments, unless they are paid by governments. This applies particularly to their role in defence and warfare. They have to remain subordinate to their governments, as they lack the legitimacy for taxation. Once a private company would start taxing people for financing public security, it would effectively function as a government.

If war is dissimilar to business, the question becomes what exactly is the relationship between war and economics. Economic factors are often the most important causes of war and they also influence the nature and duration of wars. Business has to be considered as a parasite of war. Business is not itself waging war and has no political stakes in war. It rather simply profits from the fact that there is war or its possibility, which creates business opportunities peripheral, but not central to it. In war, particularly in major war, there is the danger that private economic interests are simply overruled by politics. This is the logic of the war economies of the First and the Second World Wars in which whole nations were mobilized and their economies geared towards supporting the war effort. Although this situation still allowed private businesses to make considerable profits, their independence was curtailed and their fate became closely interlinked with the fate of the nation, in which defeat meant economic loss or even bankruptcy.

War as a business opportunity can only work in limited wars, or in conflicts which remain imaginary such as arms races. Only then can economic interests effectively dominate war or military conflict, as seems to have been the case in the late nineteenth century and to be the case today. Once the political stakes rise and/or military defeat looms, economic considerations quickly become irrelevant and the

11 Oil companies in Colombia have often their own security forces or hire local army units from the government for protecting their oil installations and personnel. Because of security concerns and increased costs some of them have left the country (Wade 2000, 32).

12 Non-state forces, which take part in war, are usually very cheap compared to regular armed forces, as they do hardly have any heavy weapons, often coerce the local population (sometimes even children) to join them, and as they tend to finance their operations through pillaging, the exploitation of readily available natural resources (diamonds), or other criminal activities (kidnapping, smuggling, drug trade). This is true for ongoing conflicts in Colombia, Uganda, Sudan, Congo, Nepal, Chechnya, and Iraq.

costs of war have to be passed on to the economy at large, quickly driving the logic of 'war as a business' *ad absurdum*.

Private companies in the defence business remain heavily dependent on governments, which are the only institutions that are legitimate actors in war and that have the financial means to finance defence. The business opportunities of private companies in realm of defence and warfare are limited by the willingness and ability of governments to pay private companies for goods and services connected to it. In democracies defence expenditures are also subject to public scrutiny and unnecessarily high defence expenditures might not be politically sustainable. The issue of cost-efficiency remains therefore important.

The Cost Saving Argument

The argument of cost-savings through outsourcing and privatization seems to be the main political justification for this development. Critics of privatization argue against this rationale and claim that privatization is not only failing to achieve cost-savings, but also could have the opposite effect. Privatization could lead to increased costs for the armed forces while causing numerous other additional problems.

There are several cost-saving factors, which can make it cheaper for the armed forces to outsource many of its non-core functions.

Higher efficiency of private companies Proponents of military privatization usually claim that private companies have to be more efficient than state organizations in order to survive on the market. This is assumed to ensure that privately provided services are both of higher quality and also cheaper than services provided by state organizations, as they lack the compulsion from the market. The industry lobbying group BENS claims privatization could do for defence what restructuring and reengineering did for American corporations like IBM, AT&T and Chrysler at the beginning of the 1990s. The problem is that legislation and state bureaucracies inhibit a necessary transformation of the defence infrastructure to a leaner and more efficient organization. BENS suggests that any non-core functions should be outsourced to the private sector and it would be possible to cut down significantly the needed defence infrastructure. As the average cost savings through outsourcing is about 10 to 30 per cent in the private sector similar cost savings could be achieved in the defence sector as well. BENS estimated the costs for the US defence infrastructure at about $160 billion (1997) and claims savings through outsourcing would be in the region of $12-16 billion annually (Taibl 1997).

Pay-as-you-use Another main argument for the cost-efficiency of military outsourcing is that the armed forces have to pay only for services and capabilities when they are needed and can save money when they are not. It does not make economic sense to have large and expensive excess capacities, which are rarely needed. Therefore, it is quite reasonable to rent services or even certain capabilities on an on-demand-basis. The best examples in this respect are transport capacities or bandwidth for satellite communications. The armed forces can retain some basic capabilities for transportation and satellite communication and hire additional capacities, if and when they are required for conducting rare large contingency operations. Pay-as-you-use might also make sense for training activities such as live training or simulator based training, as organizing such training events, as well as the maintenance of such facilities, is very costly. When it comes to international training facilities, such as Goose Bay, Canada, the costs can be shared by several armed forces, while all of them can get the benefit of it. In such sharing arrangements the overall utilization of the training facility would be much better, making it much more cost-efficient and therefore cheaper for all.

Savings on training costs By employing contractors instead of soldiers the armed forces can make substantial savings on training costs. Christopher Spearin has pointed out that basic military training costs in the US about $50,000 for each recruit (Spearin 2003, 29). It is much cheaper to hire contractors for certain specialized duties instead of training soldiers to do it, who would be in a sense unavailable for the kind of duties they were trained for, which is combat. To use a highly trained soldier for peeling potatoes is simply not cost-efficient. Similarly, it is not cost-efficient to train a soldier first as a fighter and then give him a second training as e.g. a mechanic, if one can hire a civilian mechanic in the first place. Correspondingly, the contracted companies can also save on training costs, which enables them to make cheap bids for contracts, as they are mainly recruiting former military personnel, who are often already trained for the services they are hired for.

Exotic capabilities It is sometimes not cost-efficient for the military to retain support capabilities for exotic pieces of equipment, such as the Guardrail surveillance aircraft. There is also the argument that it takes many years for the military to develop an organic support capability for a new weapons system. This is exacerbated by the growing complexity of weapon systems. It becomes then simply uneconomical. Production lines are also finite and some systems are more frequently replaced by newer systems or upgraded when better technology becomes available so that it seems to be not worth the effort for the military to always develop the competence for supporting new weapons systems (Manker and Williams 2004).

There are lots of cases when cost-savings through military outsourcing were not realized. Critics give several different explanations for this fact.

Principal-agent relations Critics of privatization often point to the Principal-Agency Theory, which analyses the relationship between client and agent in economic terms. Whenever a principal hires an agent to perform a certain function, there are the problems of different or even conflicting interests, incomplete and asymmetric information, and appropriate compensation. In order to make optimal use of an agent the principal would need complete information on the performance of the agent to compensate the agent appropriately. In reality, this is never the case and the agent has both the incentives as well as the opportunities to exploit the principal by hiding bad performance and overcharging for provided services. Performance monitoring can become a very difficult and potentially expensive task for the principal, as the principal naturally largely depends on the information which is provided by the agent. Monitoring also incurs additional costs for the principal, which could nullify the potential savings of outsourcing. Additionally, if the principal had all the competence and capability to perform a task, then the principal would not need the agent. As this is not the case, there is a structural problem which becomes even greater with the complexity of the task performed by the agent. From this perspective it seems quite unlikely that military privatization and outsourcing can save any money, as the armed forces are less and less prepared to monitor contractor performance effectively and to determine adequate compensation in the face of rapid technological change.

Lack of competition Peter Singer has argued that only competition and transparency make private services cheaper and not simply the fact that they are privately provided. Both are lacking in the defence and security business. Many contracts are awarded non-competitively and the access to the defence market by non-defence companies or newcomers is made difficult by the complicated military procurement process. Some few big defence companies command most of the procurement budgets, while more competitive smaller or non-defence companies do not as easily get contracts. Peter Singer has pointed out that many support contracts for weapons systems are long-term and this easily establishes a monopoly once the contract is signed. This practice also allows the contractor to underbid, as the contractor knows that chances are good for renegotiating the contract or for asking for expensive add-ons later. After having held a support contract for a long time the contractor can claim 'unique expertise' which influences contract rebids (Singer 2003, 155).

High transaction costs Closely connected to a lack of competition is the problem of high transaction costs for switching providers of military services from in-house to an external provider or from one external provider to another. Theoretically, the military could achieve savings by constantly switching to providers, which offer better value for money. For standardized goods and services and business relationships there are cost saving opportunities of watching the market and frequently switching providers, but even so, the costs of switching (transaction costs) need to be taken into account. Transaction costs can easily nullify potential advantages (Mowshowitz 2002, 110). Many military related services are not standardized and it is not easy to switch providers, especially if a lot of initial capital is needed for the provision of services. For example, once a company has been contracted to build and operate an expensive military training facility, it would be uneconomical to switch to another provider. Many military service contracts extend over ten or more years anyway, making switching providers very difficult and very expensive.

Bad conduct of contractors The US military uses several regulations and contract vehicles that shall ensure that it gets a fair deal from its contractors. Unfortunately, there are many contracted companies which take advantage of poor oversight and unclear regulations. This ranges from bending some regulations to outright fraud. Pratap Chatterjee describes various methods companies like KBR employ to defraud the US military in Iraq. For example, orders are chopped into several orders so that they are below the $2,500 threshold for competitive bidding. When KBR was contracted to provide such a simple thing as ice to the US forces in Iraq, the company set up a huge ice-producing plant, imported machinery for it from Texas, and employed 28 people for running the plant (Chatterjee 2004, 34). There are lots of stories of SUVs abandoned in Iraq because someone forgot a spare tire and also other forms of expensive waste, which is encouraged by cost-plus contracts (Hartung 2004). The rigging of tests for new weapons systems and all kinds of overcharging for not delivered or low quality goods and services has already been mentioned in this book on various occasions. The overall impression is that bad conduct of contractors is at least not unusual.

The Elusive Savings

Some critics like John Tepperman claim that 'virtually every time the Pentagon has bet on PMCs, it has failed to save the money it expected'. He adds 'The Defense Science Board has predicted that the Pentagon would save $6 billion from outsourcing, but this figure later turned out to be 75 per cent too high' (Tepperman 2002, 11). And Peter Singer sums up 'All too often, it [the Pentagon] outsources first and never bothers to ask questions later' (Singer 2005, 119).

In principle, there are procedures in place, which shall ensure cost-savings. The Pentagon and also the British MoD use private-public competitions as an instrument to determine which services are suitable for privatization and allow cost-savings and which are not. Military personnel 'compete' with private providers for certain functions and if the latter are cheaper and have an equal quality standard, the functions are outsourced. Ann Markusen has observed that initial estimates for cost-savings based on public-private competitions tend to be wrong, as they only 'assess the promise of savings rather than their achievement' (Markusen 2001, 8). The reality is that companies perform best during the competitions. Once they have won the contract the performance drops significantly. She quotes a Center for Naval Analysis which says a 'study of successive private contracts for maintenance of the Navy's TA-4J trainers found that the contractors (Lockheed, Burnside OTT, Grumman, UNC) performed better than the Navy in-house team in almost every case, but that for a period of around two years, contractors' initial performance was worse than that of the in-house team' (Markusen 2001, 8). Markusen's conclusion is that only routine services, which can be easily monitored by the military, tend to be suitable for cost savings, while more complex activities such as weapons system maintenance tends to be 'extremely expensive' (Markusen 2001, 16). This is a paradoxical conclusion, as it recommends outsourcing only services which can be done by the armed forces anyway, while it remains unclear how the armed forces could retain or acquire the capability of performing more complex services such as the maintenance of new high-tech weapons.

Military Privatization and Sovereignty

The most serious implications of military privatization relate to its weakening of the principles of democratic accountability and sovereignty. Arguably military privatization is different to the privatization of other public assets and services. Its long term consequence could be the re-emergence of medieval style private armies, which fight for anybody who can pay them. Military privatization could undermine the democratic control of forces, the nation state and even the international system as it exists since 1648. But how realistic is this scenario? This section discusses the political impact of military privatization with respect to the future of the state's monopoly of force and democratic accountability.

The Future of the Monopoly of Force

Different to other areas where privatization takes place such as healthcare, utilities, or public transportation, military privatization and the much more extensive role of private companies in activities relating to national and international security raises some serious questions in respect to the concept of sovereignty and the ability of states to control the use of force. Some writers already see the state wither away as a result of a broad privatization of security in which nonstate forces replace national armies as the dominant actors in warfare (e.g. van Creveld 1991, 196-197). National armed forces could be replaced by mercenary armies led by war entrepreneurs, who work in the best case for the international community as private peacekeepers or in the worst case for the highest bidder. Ultimately, corporations could control powerful military forces for attacking states or each other (Metz 2000, 21). But how realistic is such a scenario and on what kind of assumptions is it based?

The definition of the state The argument of the decline of the nation state and the rise of private armies usually rests on a narrow Weberian definition of the state. Max Weber has defined the state through its monopoly on the legitimate use of physical force within a territory. According to Max Weber, it is the *de facto* monopoly of violence that constitutes the state's legitimacy and therefore the state as such. Sovereignty means that a state can conduct its own domestic and foreign policies without outside interference. In other words, the sovereign state can determine its own fate. The guarantors of sovereignty are the armed forces which enable a state to defend itself against outside pressure and aggression. By transferring responsibilities with respect to internal and external security to the private sector the state would loose both its internal monopoly of force and its sovereignty regarding the use of force beyond its borders. The rise of private security and military companies and the increased contractor presence in contingency operations is thrown together with other forms of nonstate violence and taken as evidence that states are already powerless. This rests on the faulty assumption that it is primarily force that constitutes the state and it also confuses many rather different phenomena, which results in this flawed conclusion.

The privatization of security vs. military privatization Theorizing about nonstate violence is not easy. Robert Mandel is certainly right in his claim that the theoretical discussion of the phenomenon of a growing privatization of security in the widest sense is frustrated by a 'definitional morass' (Mandel 2002, 127). It is not the intention of this study to make it worse by adding new definitions and concepts. However, in order to discuss the wider political implications of military privatization it is essential to define and locate military privatization in the broader context of the privatization of security. The privatization of security that can be seen since the 1980s is a global and immensely complex phenomenon. It has at least two different trajectories, which have very different origins, although both have been aided by the very particular conditions present in the post-Cold War security environment (Mandel 2002. 55-57). There is a bottom up privatization of security by private illicit actors who operate effectively beyond the state. These actors challenge and undermine not only the state's monopoly of force, but also its authority and legitimacy. This includes militias, vigilantes, terrorists, mercenaries, insurgents and state terrorism. Then there is also a top down trajectory that is often initiated by the state itself and supported by its institutional and legal framework. This is about legal companies in the defence and security sectors. Table 6.1 attempts to give an overview of the full spectrum of the privatization of security.

Table 6.1 The Privatization of Security Matrix

Beyond the State — Transgression to Statehood

Organized Crime (Mafia, drug cartels)	Terrorist Organizations (nonstate, state-sponsored)	Insurgents/ Mujahedeen	Private Armies/ Warlords
State Terrorism (death squads, clandestine warfare, illegal covert action)	Traditional mercenarism (ad hoc, mainly training and combat services)	Private Military Companies (training, consulting, military intelligence)	System Contractors (operate intelligence and weapons systems)
Militias (vigilante groups)	Private Security Companies (transnational, mainly corporate security)	Military IT/ Communication Providers	Logistics Contractors (transport, overseas base operations, management)
Private Security Guards (domestic)	Facility Management (military housing, domestic base operations, depots)	Military Space (satellite launch services, operation, earth observation)	Arms Manufacturer (R&D)

Erosion of Statehood — Within the State

The matrix divides the playing field of the privatization of security in three types of actors: those who do not have to play by the rules; those who have to play by the rules, but are occasionally able to bend them; and finally those who occupy the middle ground: they nominally play by the rules, but they sometimes get into a position where they can get away with not playing by the rules. The dark squares signify the top down privatization of security, while the white squares signify the bottom up privatization of security. The privatization of military assets and the outsourcing of defence related services is clearly within the framework of the nation state. Companies in the defence and security sectors have to abide to national laws and to the authority of their domestic governments. As long as they accept the authority of their governments they can claim to be legitimate actors in the international arena. Otherwise they risk losing their main customer/client, being shut down, or even legal prosecution.

The white squares signify actors, who cannot claim to act under the authority of a state and are therefore generally illegitimate. These actors usually consider themselves as an alternative authority to the government and sometimes work towards overthrowing it. The grey squares indicate that private security and private military companies are on the one hand still within the state and subject to national and international law, but on the other hand they are already transgressing the state, its boundaries and its reach. Private security companies usually work abroad for other private actors such as multinational corporations and private military companies could (and in rare cases do) work for nonstate actors such as insurgents or transnational criminal organizations (e.g. the private security outfits that are hired by drug cartels in Colombia). They can sometimes get away with breaking the rules because they work primarily outside the jurisdiction and monitoring abilities of their home governments and because they are often too small and insignificant to appear on the radar of their home governments.

At the end of the spectrum are private armies similar to those that existed before the Westphalian system and similar to the chartered companies of colonial times like the East India Company. Private armies are different to presently existing private military companies in the crucial sense that they are not just supplementary to national forces, but would represent a complete stand-alone military force with its own assets, resources and logistical support capabilities and would be able to 'compete' with national armed forces. Under certain conditions a private military company could become a private army. Executive Outcomes came close to it during its operations in Sierra Leone (1995 to 1997). The company conducted combat operations with its own weapons and equipment (including helicopter gunships and tanks) independently of Sierra Leone's rag-tag military, which made them the strongest military power in the region. Private armies have the potential of taking over a state or carving out a state for themselves. They can also choose for ideological or financial reasons to be instrumentalized by other political actors. In Sierra Leone Executive Outcomes was in a position to take over the government, but choose not to in order to maintain its legitimacy and status as a legal corporation. Of course, Executive Outcome's presence in Sierra Leone had a huge political impact. They helped replacing the junta in charge of Sierra Leone's government with a democratically legitimized government, which was more likely to pay them (Howe 1998, 314).They also empowered a local tribe, the Kamajors, by arming them and using them in their military operations, which complicated the overall political situation in Sierra Leone (Singer 2003, 113). The existence of a private army within a state has inevitably a political impact, as it empowers private actors and weakens the state's ability to enforce its authority. But what does privatization of security mean for the principle of sovereignty?

Sovereignty and the international system Theoretically, the privatization of security could undermine the principle of sovereignty as such, once states generally lack the ability of enforcing their authority and the rule of law against private actors. Luckily, cases of a complete or partial breakdown of state authority (the so-called failed states) remain rare and are limited to the developing world. However, even a failed state remains, nominally at least, sovereign in the international system and remains entitled to the protection of its territorial integrity by the international community. A state cannot easily disappear from the map. The whole international system is based on the principle of sovereignty and states will never voluntarily give up this privilege.

States have a strong interest in maintaining their legitimacy and authority by controlling nonstate violence, including nonstate violence originating from their territory which is deployed beyond their borders, e.g. through mercenaries and private military companies. Janice Thomson has argued that monopolization of force by the state was the result of a long process of delegitimizing nonstate violence. Discussing the possibility of 'a significant erosion of the state's monopoly on the authority to deploy violence beyond its borders', she concludes that 'It is not at all clear … that this is occurring' (Thomson 1996, 149-150). Many states have a licensing system for the export of military equipment and services. That licensing systems are not toothless is proven by the demise of Executive Outcomes after South Africa's introduction of the Foreign Military Assistance Bill in 1999 (Singer 2003, 118). In fact private military companies have to be very careful not to disappoint their home governments, as their legitimacy is always in question. For this reason MPRI clears all its foreign contracts with the US State Department beforehand (Lanning 2005, 198).

Private military companies are also far too small for representing a serious threat to all, but the smallest of nations. They are definitely no match for modern national armed forces, as they cannot afford the same level of equipment and training (UK House of Commons 2002, 12). It is highly unlikely that this will change in the future. It is unlikely simply because no private company can afford maintaining forces equipped with the latest high-tech weapons. No private organization could possibly afford to purchase and operate a fleet of F/A-22s or even of modern battle tanks. Not even the manufacturers of this equipment, e.g. Lockheed Martin or BAE SYSTEMS, could afford it. Only the wealthiest nation states can pay for it. Without such expensive equipment private forces will not stand a chance in a conflict with modern national armed forces that can utilize heavy weaponry and all sorts of military high-tech.

Although Western governments have handed over an increasing amount of responsibilities to the private sector, it does not mean that they have given up the authority over the use of their forces. Private armies are not going to inherit the earth. Thierry Gongorra is probably right when he suggests that 'Fortunately, the "new mercenarism" might not expand beyond a restricted niche, namely providing military advice to small or weak states.' (Gongorra 1999, 45)

Apart from the cost factor that remains a substantial barrier to the emergence of corporate armies, air forces or navies, private actors in the area of defence and security still struggle with the issue of their legitimacy. In today's world sovereignty rests on the accepted legitimacy of the international system and not simply on the ability to use force. Private corporations lack the legitimacy to employ force without authorization and backing by states. Without this authorization they become illegitimate and they loose their ability to exist as legal entities. Sovereignty and the nation state are therefore not threatened in their existence by the top down privatization of security. It is rather the case that their sovereignty becomes transformed and more limited by the emergence of complex national and international networks of actions, institutions and regulations that intermeshes a great range of public organizations and private actors. The complexity and the dynamics of this military globalization will be explored in the final chapter. Military privatization is not the end of sovereignty, but a threat to democracy.

Democratic Accountability

Military privatization causes serious concerns about the democratic accountability of the new actors in the area of national and international security. Democratic accountability is made up of several levels of control, which shall ensure that private actors conform to democratic norms. Robert Mandel has identified six levels of democratic control that govern the use of military force and that are weakened through the privatization of security (Mandel 2002, 44). There is 1) executive control by the head of the government and by the armed forces; 2) parliamentary control through the oversight and budgeting function of the parliament; 3) public control through the power of the public in electing the members of parliament; 4) legal control through legislation and legal bodies, which governs the context and manner in which military force can be used; 5) international control that results from the need of nation states to conform to the rules and conventions of warfare; and finally 6) internal control through instilling democratic values by the way of education and training into the members of the armed forces. These mechanisms of democratic control and accountability are now undermined by the privatization of the defence and security sectors.

Ineffective executive control The control of military contractors by the government tends to be insufficient for many reasons, some of which have already been mentioned above. There are clearly problems with respect to the operational control of contractors during deployed operations, as can be seen in the case of Iraq. Nobody knows exactly how many there are and what they are supposed to do. The Pentagon has problems keeping track of all the contracts it awards and often fails to sufficiently monitor the performance of contractors. This is made worse by the tendency of Western defence departments to cut personnel. For example, while the size of awarded contracts has gone 40 per cent between 1997 and 2002, the number of Pentagon contract managers has dropped by 20 per cent (Schreier and Capirini 2005, 52).

More serious, however, is the lack of control caused by a loss of competence on part of the government. As the government keeps handing over more and more responsibilities to the private sector, it looses in the long-term its capability of making independent decisions on matters of defence. The government becomes more and more dependent on the information, expertise and advice, which is provided by provided by the private sector. In this context the notion of a *military-industrial complex* (MIC) becomes important and acquires a new meaning. The business and indeed economic survival of defence companies and other military contractors depends hugely on their ability to secure government contracts. As a result, they have a strong interest in influencing politics and political decision-makers in a way that it furthers their business. Apart from the classical MIC arguments of the iron-triangle and the political weight of defence jobs, there is a newer line of argumentation that says that the MIC's political influence increasingly rests on its edge in expertise with respect to information and technology (Gholz 2000, 38). According to this argument, the arms industry has gained more and more the ability to influence its customers because it claims to have unique knowledge on the 'security potential of new technologies' and of future threats to national security (Lovering 2000, 168). John Lovering argues that the key to the survival of defence companies is their expert knowledge on weapons technology and its uses. More important than the abilities of design and manufacturing new weapons would be their abilities 'to pull together the work of others' and to manipulate opinions in favour of new projects (Lovering 2000, 170, 174). Defence companies are not simply suppliers of goods and services in response to government demand. 'Lockheed is not just seeking to solve the problems of national security. It is framing the questions as well.' (Weiner 2004) This can lead to policies that are highly detrimental to the interests of national security, e.g. the export of arms and military services to potential adversaries, or the pursuit of militarily useless and politically dangerous defence projects such as missile defence.

Insufficient parliamentary control Military contractors working alongside armed forces in a war zone pose an additional challenge to parliamentary oversight, as media access tends to be limited and public oversight bodies have to rely mainly on information that is provided by the government/armed forces themselves. Often privatization serves as a means to avoid public scrutiny. The executive branch of the government has to account for all of its expenditures and actions and is closely monitored by legislators and the media. But by transferring more and more responsibilities in the area of security to the private sector many controversial, or even illegal government activities can be kept below the radar, which makes it easier to take controversial or illegal actions or to undermine parliamentary control.

The use of contractors in contingency operations keeps the numbers of armed forces personnel low and makes it therefore easier to deploy troops overseas. In the case of the US contractors can help circumvent Congressional caps on troop deployments. For example, Congress restricted the number of troops that could be deployed in Bosnia to 20,000. By sending 2,000 contractors more US troops could be freed for military tasks, which effectively undermined the Congressional intention of keeping the US footprint small (Castillo 2000). The use of military contractors instead of soldiers is also in other ways politically more convenient. Sometimes 'privatization' can provide 'plausible deniability' in cases where the government wants to avoid direct involvement for legal, diplomatic or publicity reasons (Yeoman 2003, 39). This could include the outsourcing of immoral or illegal research (biological warfare, mind control), illegal domestic surveillance, controversial foreign intelligence operations (assassinating, detaining or torturing alleged terrorists) and could go as far as the secret intervention in ongoing conflicts with mercenary forces, e.g. Angola, Nicaragua, Colombia and Sierra Leone.

Weakened public control Privatization reduces public oversight, as private companies are less required to lay open the details of their business operations. Peter Singer has pointed out that contracts are often treated as proprietary and are not subject to public scrutiny (Singer 2004d, 17). Public bodies in the US, the UK and Germany have to disclose information under the Freedom of Information Act, but no similar obligation exists for private companies. When things go wrong, as in the case of Northrop Grumman contractors whose plane crashed in rebel-held territory in Colombia in 2003, the company could and did reject giving information to the public about the nature of its business in Colombia (Schwartz 2003, 100). Private companies are also mainly accountable to their owners, for instance their shareholders, but not to the public at large. They have a legitimate interest of withholding information that affects their competitiveness, which clashes with the requirement of transparency necessary for democratic accountability. Big corporations only have to publish financial information and very general information on their business operations in annual reports. However, smaller and not publicly traded companies usually do not even have this obligation. They can refuse to publish any information on their business.

Even if there is a direct involvement in a conflict, it is more politically convenient to substitute soldiers with contractors. The death of soldiers in action has to be reported by the government and this usually receives a lot of media attention, which affects public opinion. On the other hand, contractors killed in action hardly make it into the news and it is also far less troublesome for the government. There is no official monitoring and there are few reliable statistics and estimates on contractor casualties. It is believed that about 800 US military contractors died in Iraq and that 3,300 have been wounded (Roelofs 2007), but nobody knows the exact numbers. Exposing contractors to the dangers of a war zone is cheaper for the government and also cheaper for the contracted private companies in case they become casualties. When soldiers are wounded or killed, the government has the responsibility to take care of them or pay a life insurance to their relatives. If contractors are wounded or killed, it is essentially their own risk and their own problem (Gegax 2004). Only recently have the families of killed contractors taken the step of suing the security company Blackwater USA, which has supplied thousands of armed guards in Iraq, for negligence. Blackwater is still fighting these suits and does demand similar immunity from legal action as the armed forces enjoy (Roelofs 2007). The future of the industry will very much depend on how it will be regulated.

Ineffective legal control Military contractors, who show poor performance, who break national or international law, or who generally act in opposition to the interests of their home countries, should be penalized. Unfortunately, this is rarely the case. A major obstacle is that poor performance, fraud and other criminal behaviour remains undetected because of ineffective control and insufficient oversight. But even if misbehaviour on part of military contractors is exposed by the government or the media, the reality is that there are often few means of holding them accountable. In theory, military contractors who fail to perform or break the law, can be excluded from future contracts or tried in court. In reality, the concentration in the defence sector makes it increasingly difficult to switch providers or to take cases to court. For example, it was quite exceptional that the US government decided to prosecute Boeing for spying on its competitor Lockheed Martin in 2003. This led to the suspension of three of Boeing's business units and three of Boeing's executives (Zucker 2004). However, it is unlikely that Boeing could be forced to leave the defence business, as the US forces already depend far too much on Boeing with respect to procurement and operations. A concentration of military experience and knowledge in the form of the military consulting company MPRI could soon be just as difficult to replace as a major defence manufacturer like Boeing.

Sometimes the government might not even have any interest in prosecuting a military contractor, especially if this would hurt economic interests or conflicts with foreign policy objectives. For example, there was little interest from the side of the British government to prosecute Sandline International for its breach of the UN arms embargo to Sierra Leone in 1998, as British government officials were involved in it. Similarly, the British government decided to stop an investigation of BAE SYSTEM's bribery of Saudi officials in connection to the £6 billion Eurofighter export deal to Saudi Arabia in 2006 because of the importance of the deal for BAE.

Weakened international and internal control It has already been discussed above that international law is outdated and that military and security contractors often operate in a legal no-man's land. National armed forces have a clear obligation to conform to international law and to prosecute breaches of the conventions of war. These norms are universally accepted and firmly established. However, even Western armed forces occasionally fail to adhere to the standards of international law and usually armed forces have little incentive to prosecute its own members for human rights abuses in wartime, as this weakens public morale. It is debatable if military and security contractors have more or less incentives to conform to international law. On the one hand, they cannot afford the image of being mercenaries on a rampage, as this would get them out of business, as former Executive Outcomes chief van den Bergh claimed (Arnold 1998, 119). It is also true that most military and security contractors have served in national armed forces. They are therefore familiar with the conventions of war and accept them as intrinsic values. On the other hand, private companies can often evade monitoring by oversight bodies or the public and they have even less incentives than the armed forces to bring its members to justice for abuses. In addition, private military companies often have insufficient vetting procedures, which allows war criminals to join them. In the long term the pool of 'ex' Western armed forces members will become smaller and as a result they will increasingly be forced to recruit less trained individuals or people from developing countries, who might not have the same self-restraint as the members of Western armed forces.

Conclusion

This chapter has looked at the military, economic and political implications of military privatization. The phenomenon of military privatization and its implications are very complex. It is hardly possible to give a straightforward answer in respect to the question whether there are more advantages or disadvantages to it. From the point of view of the armed forces, military privatization threatens the unity of command, causes new vulnerabilities and leaves military and contractor personnel with a high degree of legal uncertainty. But there are also military advantages of using contractor instead of military personnel. It allows the military to utilize technologically very advanced weapons systems, which indeed have greater capabilities than less advanced weapons. So there is a trade-off between contractor-enabled high-tech systems and less advanced systems that can be operated and maintained by the armed forces themselves, which needs to be taken into account. High-tech weapons are not always militarily decisive, but for certain types of high-intensity warfare they probably are. However, the most worrying military implication is the general loss of competence by the armed forces, which could make it very difficult to politically reverse the current trend of privatization.

Similarly there are certainly economic advantages and disadvantages of military privatization. In the analysis it has been established that the use of armed force is in itself not a profitable activity and is therefore subject to political objectives. Defence equipment is not freely tradable and defence markets are limited and depend largely on the overall political climate. War is indeed very different from business and it can only be a good business opportunity under specific circumstances, for example in conflicts, which remain limited or imaginary. It has been claimed that military privatization makes at least in peacetime good economic sense, as military privatization can offer the promise of savings in defence expenditures. On the other hand, analysts have often pointed out that in reality savings were always far smaller than expected.

Finally, the political implications of military privatization have been analysed. In respect to a possible erosion of the principle of sovereignty, it seems at the present quite unlikely that states will lose the authority over the legitimate use of force. There are two trajectories of the privatization of security, which need to be distinguished: the bottom up privatization of security and the top down privatization of security. While it remains unclear whether illicit private authority could replace the authority of states, it seems evident that top down privatization of security in the form military privatization is not going to threaten the survival of the nation states or the international system. However, there are very serious implications for democracy, as democratic control mechanism of the use of force have been substantially weakened. This encourages corruption, political risk taking and the pursuit of unwise or even dangerous policies. The final chapter will look at the growing complexity of national and international security arrangements and relationships that is the result of the ongoing military privatization and globalization.

Chapter 7

Military Globalization

The literature on globalization is still rapidly growing and, despite disagreements in regard to extent and consequences of the present globalization, there seems to be some consensus on its meaning. Globalization has been defined as 'the process of increasing interconnectedness between societies such that events in one part of the world more and more have effects on peoples and societies far away' (Baylis and Smith 2001, 64). This interconnectedness is caused by an 'expansion of cross-border networks and flows' (Tangredi 2003, 301) of people, goods, capital and information. Some see globalization as an all-encompassing process resulting in a politically, economically, and culturally unified world. Globalization also stands for an emerging (economic) world system, which connects all world regions and integrates them into one world economy (Wallerstein 1979). On the other hand, one needs to be very cautious seeing globalization as just one coherent phenomenon. Instead, globalization should be seen as a complex of different overlapping and sometimes diverging processes, which contest, influence and possibly reinforce each other. Or in other words, the global system is an open system governed by lots of positive and negative feedback loops with 'remarkably different and contrasting trajectories' (Urry 2003, 102).

The process of increasing extent and scope of security relationships has been called by David Held 'military globalization'. He distinguishes 'military globalization' from 'global militarization', which would be synonymous for a global military build-up.[1] Military globalization, on the other hand, does not imply that the world gets militarized in a global arms race, but rather stands for the firmer integration of armed forces around the world into the global military system. Military globalization is defined as 'a process which embodies the growing extensity and intensity of military relations among the political units of the world system … Understood as such, it reflects both the expanding network of worldwide military ties and relations, as well as the impact of key military technological innovations (from steamships to satellites), which over time, have reconstituted the world into a single geostrategic space.' (Held e.a. 1999, 88) Held divides military globalization into three distinct phenomena: 1) the globalization of the war system; 2) the global system of arms production and transfers; 3) the geogovernance of violence. The first point refers to the 'geopolitical order, great power rivalry, conflict and security relations'; the second to the global arms dynamics; and the third to 'embracing the formal and

1 Keith Krause uses the term 'global militarization' to explain the diffusion of sophisticated weapons systems and the tendency of states in conflict-prone regions such as the Middle East to stockpile these weapons (Krause 1992, 188).

informal international regulation of the acquisition, deployment and use of military force' (Held e.a. 1999, 89).

Like globalization as a whole, military globalization is not a coherent process leading to an in the long term inevitable result, but itself a phenomenon with several different layers, processes and inherent contradictions. For example, it does not appear that the world is becoming militarized, but rather that a global civil society has emerged, which marginalizes the role and importance of military force. While Western societies have become post-military, their armed forces have also become civilianized. Military globalization and civilianization are contradictory, but also mutually reinforcing processes. All three processes of military globalization distinguished by Held are connected to technological development, which made them possible in the first place. The result is increasing global interdependence and complexity.

The Globalization of Military Activities

How did the process of military globalization start and what does it mean for security relationships between states and the international system? With the Age of Discovery, the colonial expansionism of Western states, European militaries began operating globally in the sixteenth century. The 'fire-power gap' allowed Britain to annex one third of the world's land mass by the end of the nineteenth century. Other European colonial powers were similarly successful. The imperial rivalry led to the First World War, which was the first global conflict in world history. The transportation revolution of the nineteenth century had even before increased the speed and reach of military operations tremendously. The Second World War saw the emergence of global weaponry in the form of ballistic missiles and long-range bombers (Baylis and Smith 2001, 16). The Cold War was a quest for global domination between the superpower blocs and expanded the military activities of the two superpowers to virtually every corner of the world including the deep sea and space. No point on the surface of the earth is now in principle out of reach and no target can be sufficiently hardened to escape destruction.

The 1991 Gulf War was in many respects a globalized conflict, although the actual fighting was limited to Kuwait and Iraq. It was globalized because support and resources for this war were mobilized from all over the world. The coalition forces were composed of soldiers from 28 nations; they and their equipment, including many supplies, had to be carried half way around the world; an intercontinental bombing raid was conducted from over 7,000 miles away at the outset of the war; the coalition soldiers used global assets such as satellites; the war itself was financed by various nations from all over the world including Germany, Japan, Saudi Arabia and Brunei; and the war was presented to a world audience in real-time on TV (Baylis and Smith 2001, 16).

Technology has proliferated around the globe and has enabled more nations to operate militarily on a global scale, often as part of peacekeeping forces or of coalitions. The operational tempo for Western armed forces has increased substantially since the end of the Cold War, despite the fact that force levels are decreasing. This was only possible because of the availability of a predominantly civilian global infrastructure. Nevertheless, Western armed forces are continuously improving their own force projection capabilities. In the 1990s the global military communications networks have been updated and even middle range military powers such as Britain and France are already using their space assets for supporting their global military activities. Many other nations have acquired access to space assets, or make use of commercial assets. Globalization creates both the possibility for global military activities, but also the need for them (Tangredi 2003, 299-315). According to Sam Tangredi, armed forces have to fight transnational and non-state threats, which means that they need to go wherever global terrorists and criminals have found a safe haven; secondly, they need to protect global trade and in particular the sea lanes, as the global economy relies increasingly on shipping; thirdly, economic security demands the safeguarding of the civilian infrastructures against terrorism and crime; and finally, greater importance of international security arrangements and commitments make it more likely that armed forces take part in international operations (peacekeeping, humanitarian or military interventions) (Tangredi 2003, 302).

Military Globalization and 'Complex Interdependence'

Globalization has created a situation, which might be best described by Robert S. Keohane's and Joseph S. Nye's term of 'complex interdependence' (Keohane and Nye 1998, 81-94). Political, economic, and cultural relations have increased in scope, extent and intensity. These intensifying relations of states, societies, and non-state organizations tend to be reciprocal and can lead to mutual dependence. In other words, the actions of states can impact upon other states and the consequences of the actions can also affect their originator in sometimes unforeseen ways. Interdependence can be symmetric, which means that actions produce an equal impact on actors, or they can be asymmetric, which means that some actors are affected more than others. Interdependent relations can be described in terms of 'sensitivity' and 'vulnerability'. Sensitivity is defined as 'the degree of responsiveness within a policy framework – how quickly do changes in one country bring costly changes in another' and vulnerability as 'the relative availability and costliness of the alternatives that various actors face' (Keohane and Nye 1989, 12). Kenneth Rogerson claims that complex interdependence limits the freedom of a state, as it 'relinquishes some ability to control its own activities'. This could mean that 'sovereignty can decrease' (Rogerson 2000, 421).

Complex interdependence also characterizes security relationships between states. Security is still a main concern for states and the use of force still plays a role in international relations (Keohane and Nye 1998, 84). States are no longer self-sufficient in respect to security and global defence cooperation is more important than ever (Roosevelt 2005, 1). This has many reasons and one that goes back to the First World War is the realization that alliances are needed in a global confrontation. States have increasing difficulties estimating the true capabilities of adversaries, as weapons incorporate so much knowledge that is implicit and hard to measure; as additional capabilities can be quickly acquired on global markets; and as military technology can be easily transferred, or hidden in form of blueprints or virtual arsenals. This weakens deterrence and makes arrangements of collective security correspondingly more important. Secondly, the costs of security have increased to an extent that they are hardly affordable. No nation can afford to possess all the different capabilities necessary to prevail in all kinds of conflict. The spectrum of warfare has become bigger and newer types of military capabilities and operations have become much more expensive. Thirdly, democratic states are increasingly under pressure to spend money on welfare rather than defence. Finally, the nature of the security threats has changed in a way that it is practically impossible for any state to deal with them effectively without international collaboration (illegal immigration, transnational organized crime, global terrorism, weapons proliferation, the missile threat).

Dealing with contemporary and future threats to international security will require a lot of international cooperation, not only militarily, but maybe more importantly, diplomatically, economically and technologically. For example, failed or failing states are perceived as a threat to international security, as they allow terrorists and transnational criminal organizations to use them as safe havens and bases of operations. These non-state actors do not only further destabilize failing nations, but can pose a security threat to the rest of the world as well through executing terrorist attacks, or the trafficking of drugs, weapons and people. Western states sometimes need to prop up governments in failing states so that they can regain control over the whole of their territory and deny terrorists and criminals a safe haven. Sometimes it will be necessary to militarily intervene in failed or failing states and to engage in nation-building.

As terrorists and criminals can easily shift their operations from one country to another and as threats to security can suddenly appear in form of WMD proliferation, it will be necessary to very quickly assemble coalitions. Future coalitions are very likely to be ad hoc and there will be a whole range of factors influencing states in their decision of taking part in a coalition operation, or not, e.g. geography, financial aspects, available capabilities and interoperability. Additionally, the right mix of capabilities needs to be assembled through coalitions for each particular operation. The result is what Alvin Toffler sees as an important aspect of future warfare, which is the growing reliance on 'modular coalitions as crises arise' or which Jan Nederveen Pieterse calls 'temporary plug-in/ plug-out coalitions' (Toffler and Toffler 1993, 98; Nederveen Pieterse 2002, 75-91). The War on Terror has already led to the assembling of such temporary coalitions and more of such can be expected.

Consequences of the Global Arms Dynamics

The emergence of a global arms and security market has many important implications for national and international security and the conduct of war. The present world military order needs to be understood as a complex system in which the global arms dynamics plays a crucial role. It is assumed that arms remain central to the political world order, warfare and the war system.

The Global Arms Production and Transfer Network

Arms production and transfer is not a one-way street. Keith Krause has pointed out that 'The global arms transfer and production system involves more than just the producers of arms: it enmeshes all major and most minor states as patrons and clients in a complex web of military relationships that includes not only weapons, but arms production technology, spare parts and supplies for weapons, and military training assistance.' (Krause 1992, 182)

Complex weapons systems such as fighter jets, which have proliferated quickly around the globe during the 1980s and 1990s, require by their very nature a long-standing relationship between suppliers and recipients. Recipients need in addition to the weapons systems themselves, training and continued access to spare parts, as well as maintenance, repair, overhaul and upgrade services, which keep the hardware operational in the long term. Often such sales are combined with offset agreements of coproduction or licensed production of weapons systems and spare and wear parts. It is increasingly the case that the supplier also becomes economically dependent on recipients and that recipients gain more than just arms, but also know how, which reduces the dependency on the supplier. As there is now a global arms market that is no longer constrained by ideological alignment, it is not difficult for less developed states to purchase weapons from various Western or Eastern suppliers. Many states without a significant defence-industrial base already use a mix of Western and Eastern equipment.

The proliferation of high-tech weaponry increases the possibilities as well as the need for military cooperation. First of all, offering military training to foreign armed forces is usually seen as a method of initiating arms sales. For example, the US as the world's greatest arms exporter has an ongoing military cooperation with most countries of the world. Lora Lumpe claims that the US trains annually 100,000 foreign soldiers in 180 countries (Lumpe 2002, i). Also other major arms exporters like Britain, France and Germany use the military training of foreign soldiers as a marketing instrument. Secondly, once a country has purchased foreign equipment, it might become dependent on foreign support for operating it. If support through the exporting nation can no longer be secured, it might be possible to get support from other nations or other suppliers. The more widespread the equipment is, the more likely it is that other support options can be found.[2]

The Diffusion of Military Technology

As the American/Western military superiority rests on the use of more advanced technology, the diffusion of technology can threaten this superiority (Paarlberg 2004, 122-151). There are basically three main causes for the diffusion of military technology: 1) arms exports; 2) dual-use products; and 3) technology transfer.

---

2 For example, Indonesia has purchased F-16s from the US in 1990, but as the US has withdrawn its support because of an arms embargo since 1999, the airworthiness of these fighter jets has become questionable. Now Turkey has offered military support for the F-16s to Indonesia in an attempt to improve the ties between the two nations (Dewitte 2006).

The trade of arms and military services is governed by international and national regulations. There are international weapons control regimes like the Missile Technology Control Regime, the Nuclear Non-Proliferation Regime and the Wassenaar Arrangement on Export Controls for Conventional Arms and Dual-Use Goods and Technologies, which are intended to prevent the proliferation of weapons technology. Additionally, there are national arms export restrictions like the American International Trade in Arms Regulations (ITAR), which restrict US companies with respect to which items and services they may export and to which countries. Sales of military and dual-use products also require an end-user license that prohibits the recipients to sell these products or the technology that they incorporate to third parties. However, in the post Cold War world some of the restrictions have been relaxed for economic reasons and the international arms trade is well and alive. Arms export statistics only give the value of arms exports, but say nothing about the sophistication of the weaponry. While the value of arms transfers remained quite stable over the last decade, more advanced weapons have been exported. The world's biggest arms exporters have exported some of their most advanced weapons to allies even before they entered service in their own armed forces. Some of Russia's most advanced weapons are exclusively produced for export, as the Russian armed forces cannot afford them (Blank 1997, V). Russia was the biggest arms exporter in 2004 and most of the weapons went to China, the largest recipient in the same year. China has purchased many of Russia's most advanced weapons including the BTR-80 and BMP-3 (APCs), the S300 PMU (SAM), T-80 and T-90 (MBTs), the Su-27, Su-30MK, Su-35, MiG-29 and MiG-31 fighter jets, the Tu-22 bomber, Kilo Class submarines and Sovremny II Class destroyers (Sergounin and Subhotin 1999, 70-83). Also other nations in South and South East Asia were recipients of some of these weapons systems including India, Myanmar, Laos, Singapore, Malaysia, Indonesia, Vietnam and Thailand. Similarly, the Americans, British, French and Germans have exported some of their newest hardware. For example, the Americans exported the F-14, F-15, F-16, F-18 fighter/bomber series also in their newest versions, the Apache and Blackhawk helicopters, the M1A1 (MBT), the Bradley (IFV) and the Predator UAV. The Europeans exported the Eurofighter Typhoon, the Super Lynx helicopter, the Rafale, the Leopard 2A5, the Pz2000 howitzer and so on (SIPRI 2005, 455-521). Many of these weapons ended up in the rich, but politically unstable countries in the Middle East and South/East Asia. In some cases poorer countries decided to lease main battle tanks, F-16 jets and even warships (Matthews and Parker 1999, 27). Export restrictions alone do not seem to be a serious obstacle for many states in the developing world acquiring sophisticated weaponry from either the West or Russia/China. Often arms exports are justified by exactly this argument of 'if not we, somebody else will sell them'.

Even if countries are not able to purchase high-tech weapons because of export restrictions, the availability of dual-use items combined with some foreign know how can enable them to put together their own equipment. Dual-use technologies are by far more difficult to control than genuine weapons. Nevertheless, there are controls such as the American ITAR, which covers the licensing of dual-use goods, the European Union Dual-Use Directive and the Wassenaar Arrangement, which specify which goods require an export license. The problem is that it is very difficult to identify, which commercial goods and technologies are defence relevant and should not be traded. Another problem is that governments disagree widely in respect to whether a commercial item has to be classified as dual-use item and should be restricted and, if it is classified as such, how these restrictions should be handled. Most of the technology utilized for the RMA comes from the commercial world anyway. As dual-use products have become increasingly important for military procurement of Western armed forces, they are theoretically also available for potential enemies. These countries can get access to important capabilities such as satellite communication/navigation and earth observation through commercial sources. Third World countries can additionally acquire relatively easily very sophisticated weaponry by purchasing commercial components and the knowledge for integrating them into weapons systems on the global market. For example, Pakistan could this way recently build its Babur cruise missile with easily available commercial components and some help from China (Hewson 2005, 41). It has been estimated that there are already 80 countries, which have cruise missiles of which 18 are able to produce them domestically (Bolkcom and Squassoni 2002). It is anticipated that the number of countries with cruise missiles, including UAVs and Land Attack Cruise Missiles, will continue to grow, as export restrictions are undermined by the sale of commercial components used in civilian aircraft, which can be converted and put to military use. The proliferation of missile technology to Third World countries poses a significant threat to Western militaries wanting to operate near, or intervene in those countries. In this respect a lot of lessons could be learned from the Falklands War, which the British almost lost because of the threat posed by the French-supplied Argentinean Exocet missiles (West 1998).

Knowledge transfer through collaborative R&D and production of military equipment is another important reason for the diffusion of technology. Governments generally try restricting the foreign access to technology. This obviously includes the issue of foreign ownership and control of defence and other technology companies. Foreign takeover of domestic defence companies are subject to government approval, which is rarely granted, especially in the US. The US market is largely closed to European defence companies. The only exception are British companies, which already own significant parts of the US defence industrial base. But even in this case, the US branches of British defence companies remain organizationally separate and do usually not transfer technology amongst each other. There are international joint ventures for defence projects, especially within Europe (e.g. Eurofighter, Eurocopter, A400M), which usually implies transfer of knowledge. However, this is limited to particular projects and close allies. On the other hand, in offset arrangements the extent to which technology is transferred is difficult to control (McLean 2005, 6-7). Military offset agreements often include knowledge-transfer in some form. When high-tech weapons are exported the recipients sometimes acquire the right to produce part of the weapons or some of their components under licence. In the long run this could mean that recipients can acquire the knowledge to produce some sophisticated weaponry themselves. China is already able to design and build its own weapons such as fighter jets and missiles. Much of the technology came from Russia through recruitment and exchanges of Russian scientists and personnel (Sergounin and Subhotin 1999, 83-89). Apart from government to government knowledge transfer, a lot of technical expertise can be purchased from private companies on the global market, for example systems integration services, military training and consulting, all of which contributes to the diffusion of military technology and knowledge.

There are still some important barriers to the diffusion of military technology and knowledge and to the globalization of the defence industry. This includes international arms control regimes, national licensing systems and national restrictions in respect to the ownership and control of defence companies. Despite of this, the defence industry is becoming more globalized and defence-relevant technology is becoming more accessible around the world. With the globalization of the defence industries, the easy availability of sophisticated dual-use goods and the new emphasis on arms exports for economic reasons, it is almost impossible to prevent the proliferation of technology. However, access to military high-tech does not imply that less advanced countries can master the technology, or could avoid dependence on foreign support and supplies.

Military Hegemony and Complexity

The most striking feature of the present situation is the exceptional military predominance of the US. The US not only accounts for almost half of the world's military expenditures (SIPRI 2005) and maintains a global network of 700 military bases and installations abroad (Johnson 2004b), but is also by far the world's biggest arms producer and exporter.[3] America 'controls the development, production, and sales of the world's most sophisticated weapons, especially precision-guided arms, information systems, and communications' (Harkavy and Neuman 2001, 301). It spends several times more on military R&D than all European countries combined and there is little reason to assume that the American technological lead will get any smaller in the medium term (Paarlberg 2004, 122-151). It is even likely that the technological gap between the US and its European allies is widening with possible consequences for the alliance.[4] However, this dominant position does not mean that the US is less exposed to the overall dynamics of military globalization. In fact the US is as much driven by and vulnerable to the global arms dynamics, as it is able benefit from it.

---

3 According to SIPRI 38 American companies and one Canadian company account for more than 63 per cent of worldwide arms production (SIPRI Yearbook 2005: Chapter 9) and according to estimates by the US Congress for the period 2000-2003 the US accounted for $76 billion out of total $148 billion in arms transfers or about 51 per cent (Korb and Wadhams 2005).

4 A study by the National Defense University claims that 'with U.S. C4I capabilities advancing rapidly through accelerated investment in transformational technologies, the interoperability gap with fielded U.S. systems is likely to increase, not improve, in the years to come.' This could mean that the newly created NATO Response Force (NRF) is 'dumbed down' with significantly inferior capabilities to the US forces. As it is questionable 'whether any country would be willing to put its forces at risk under these circumstances', the NRF might not be usable for high-intensity conflicts (Bialos and Koehl 2005, 78).

Network-Effects of Military Networks

The most modern armed forces are aiming for a 'transformation', in disregard to the known problems with military high-tech, as well as its relative irrelevance for many kinds of conflicts.[5] The point of the RMA is that it makes warfare more capital-intensive, as it encourages militaries to develop or buy weapons and equipment at the technological frontier. 'Smart' weapons remain several times more expensive than ordinary munitions, while 'brilliant' munitions that can independently acquire targets are more expensive still (Silverstein 2000, 24). Keith Hayward has pointed out that 'RMA technologies are extremely complicated to master, test, produce, deploy, assimilate and maintain.' (Hayward 2000, 125) It is unlikely that a transformation to network-centric warfare (NCW) would be possible on the cheap, thus limiting the number of countries able to participate in this technological arms race. As a result there is a growing military divide between nations that can compete on a high level and those which cannot (BICC 2004, 28).

The American pursuit of the RMA has already caused some serious concerns and friction within NATO. The main issue is interoperability. The whole idea of NCW is that the armed forces can function as one unit, allowing completely new tactics. Coalition forces which are not fully integrated into the network pose a problem, if not a hindrance, or even a risk. If the European militaries are no longer interoperable with the American forces because of their outdated equipment, they will not be able to contribute significantly to future coalition operations. This way the Europeans are practically forced to modernize their militaries, if they do not want to become militarily irrelevant. There is additionally the fear that the US could dominate also the European arms market. Europe has for this reason pushed for the consolidation of its defence industries, as the growing competition with the US threatens the survival of an independent European defence industrial base (Wilson III 2005, 84). The present modernization projects of European armed forces can only be understood within the context of the American RMA and American defence spending. A general modernization of NATO forces has also encouraged the possible American peer competitors Russia and China to modernize their forces as well.

In respect to network-centric systems there is clearly a self-reinforcing mechanism at work. The Americans organize their armed forces as a network and ostensibly gain a military advantage by doing so, which was already to some degree demonstrated in the 2003 Iraq War (Fein 2005). Future allies and coalition partners are now called upon to modernize in order to remain interoperable with US forces. However, the more widespread military networks become and the more compatibility is achieved through international technical standards, the more useful they become as well. Interoperability not only with US forces would be achieved through the introduction of compatible network-centric systems, but with many other armed forces as well. Coalition forces could be easily plugged into the network, making ad-hoc coalitions

---

5 As the world's population increasing lives in big cities, future military operations will most likely take place in urban environments. Alice Hills claims that urban operations are distinctive and that technology is often not an answer to the particular challenges posed by urban environments (Hills 2004, 9).

and coalition command and control a lot easier. This creates powerful incentives for developing or buying network-centric systems. The most advanced military equipment will have the capability to connect to military networks and the more demand for it exists, the more of it will be available. Raytheon estimates that $240 billion of equipment oriented towards NCW will be procured over the next ten years (Harz 2005).

Towards an American Arms Hegemony?

The US is already the world's main supplier of weapons and the US also uses arms exports to influence recipient nations. The sometimes reciprocal dependencies of suppliers and recipients of arms are known and also the tendency of supplier nations to act as patrons to the recipients. Although dependencies in arms transfers tend to be asymmetric, they are not one-sided. Suppliers also become economically dependent on arms sales and can sometimes be drawn into conflicts by their clients. Recipient states also try to reduce their dependence on a particular supplier by purchasing equipment from many different sources. They can also try to develop their own weapons based on designs taken from imported ones. As a result, the leverage of a weapons supplier over its client can be very limited.

However, the RMA has added new dynamics to weapons transfers. As there is a widening technological gap between the US and the rest of the world, it gives the US not only a military advantage, but also an economic one. Nations aspiring to have the latest military technology, or wanting to be part of a US led coalition, might feel compelled to buy American equipment. This allows a new approach to defence exports best symbolized in the JSF (Context 2003). The US can bully its allies to purchase its products so that interoperability is guaranteed. If Lockheed Martin can deliver the JSF at the projected price of $50 million, the plane could dominate the world's military aircraft market and destroy the European military aircraft industry. It is partially a matter of sheer volume: the US and Britain plan to purchase altogether 3,000 aircraft and hope to export another 3,000. This compares with just 1,000 Eurofighter, which are planned to be produced by 2014 (Fallows 2002, 62-74). Already eight US allies have signed up for the JSF.[6]

The Americans also want to keep aspects of the technology of the JSF secret making it difficult for their allies to do maintenance and repairs on the fighter jets themselves. Antonia Feuchtwanger quotes a British defence source saying that: 'If the Joint Strike Fighters get a single bullet hole in the tail, we do not have the know how to mend the stealth covering. We'd have to get to the back of the queue behind the US Marines, US Army, and US Navy fighters.' (Feuchtwanger 2004, 56) The foreign buyers of the JSF would be dependent on Lockheed Martin's support capacities, which could mean unacceptably long waiting times, but also that the Americans could withdraw their support to allies completely, leaving them with unserviceable aircraft.

The US can exploit its dominant role as most advanced arms producer and biggest arms exporter in another way too. As the world becomes more dependent on

6 Britain, Italy, Netherlands, Turkey, Australia, Norway, Denmark, and Canada.

American arms, it makes itself vulnerable to unpleasant surprises. It is conceivable that exported high-tech weapons, which contain a lot of complex software, could simply be switched off by the manufacturer. The software of exported weapons systems could contain hidden functions, which limit their uses. For example, as weapons systems such as fighter jets utilize GPS, it would be possible to restrict the use of an exported fighter jet to a certain geographical area. If the jet leaves that area, the plane might explode, or the pilot might be ejected (Toffler and Toffler 1993, 278).

Not without reason do the British want the source code for the Joint Strike Fighter from the Americans. They have concerns that the Americans could be able to switch off the fighter jets whenever they wish (Chapman 2006). The possibility of inserting hidden safety switches into exported high-tech weapons systems already puts some strain on military alliances. Some countries therefore demand the source code for any weapons they buy, which poses a dilemma for the seller nation in terms of technology proliferation and competition. Even if the source code is provided, software used for weapons systems can be so complex that it would not be easy to discover hidden functions. For example, the source code for the JSF might have 19 million lines (Tewes 2006).

It can be expected that it will be impossible in the near future to sell any high-tech weaponry that cannot connect to military networks and make use of satellite communications and navigation.[7] Few nations have their own space capabilities and even fewer have their own satellite navigation systems. In fact there are presently only two systems in existence: the American GPS and the Russian GLONASS. The European system Galileo will only be deployed by 2010. Satellite-guided weapons have been one of the best-selling military export articles since the Iraq War (Merle 2003). Any nation that buys satellite-guided weapons would have to use one of the aforementioned three systems. The nation, which operates the navigation satellites, could easily switch off or jam sold satellite guided munitions, thus making the weapons useless for the buyers. The only way for a nation to get around this problem would be to put its own GPS into orbit, which is very expensive. However, some nations might be able to afford it. For example, the Chinese are already planning their own global satellite navigation system called Beidou.

The Complexity Challenge

The most modern armed forces have to cope with ever increasing technological, organizational and environmental complexity. This complexity creates innumerable dangers for them. It is not just a question of whether complex technical systems are performing sufficiently well in real contingencies. Equally important is that governments have lost their capability of effective control over military relevant technologies, as the private sector plays the crucial role in developing them. The economic pressures for exporting defence equipment and services, as well as

7 There is already an attempt to establish international industry standards for network-centric systems so that they would be compatible to each other and to allow different platforms to be more easily integrated into networks.

the international sharing of defence assets such as satellites, can make security relationships very complex and can create a potential for heightened tensions and conflicts.

It will also be more and more difficult for governments to check up on the activities of globally operating and interconnected defence companies. In some cases defence companies might even refuse to cooperate with their home countries. For example, Dassault protested against Mitterand's demand to give the Exocet codes to the British for sabotaging the missiles during the Falklands War (Follain 2005). There is no guarantee that they would in all cases oblige to the wishes of their home governments, especially if this would hurt their business interests.

Military globalization also creates lots of new vulnerabilities for technologically advanced states or First Tier weapons producers. Firstly, they are no longer in control of the whole supply chain for the production of weapons. Many components come from abroad, which means that these states could experience shortages of crucial parts in times of conflict and war. Secondly, it will be increasingly difficult to look through the complex global networks through which military high-tech is created. For example, with the growing importance of commercial software embedded in military systems, it 'could come from anywhere in a global industry' (Hayward 2000, 118). Because of the ever-increasing complexity of software malicious programmes would be extremely difficult to detect, and could facilitate system intrusion or otherwise impede the functionality of military systems (US DoD 1999, 18). Thirdly, with the growing dependence of armed forces on commercial technology and civilian infrastructure, it is a lot easier for adversaries to penetrate military systems and to disrupt them. Paradoxically, this means the more advanced weaponry a nation uses, the more vulnerable it becomes to information warfare. Finally, the growing role of global defence companies for supporting military operations creates vulnerabilities and uncertainties for the armed forces, which have to rely on them. Will these companies provide the services when they are needed most, or will they pull out for whatever reason and degrade military capabilities? Military contractors also form the soft underbelly of the most advanced militaries. They become the most likely targets for asymmetric attacks in theatre and even at home. It could be possible to disrupt military operations by attacking military suppliers and supply chains. For example, Lockheed Martin's computers were attacked by a 'worm' in August 2003, forcing the defence giant to shut down part of its network (Dowd 2003).

There is the danger that the US and the West in general will not be able to master the growing complexity of technology and security and that the Western armed forces are moving towards criticality. The utilization of ever more complex weapons increases vulnerabilities, which need to be addressed. As Chalmers Johnson points out this might lead to an endless process in which the US has to expand its power more and more until it overreaches and the costs of continuous expansion can no longer be financed. For example, the US must control outer space because it is critically dependent on space-based systems, meaning it has to control the access to space, meaning it has to control the technology that allows nations the access to space and so on. Johnson compares it to Britain's need to protect the sea routes to India, which required the control of Egypt and South Africa. To protect Egypt, it had to control the Upper Nile, hence the expansion into Sudan. To protect South Africa, it had to fight the Boers. It is according to Johnson, like the 'domino theory': the American logic of protecting its military advantage 'leads to an endless progression of places and commitments that must be protected, resulting inevitably in imperial overstretch, bankruptcy, and popular disaffection' (Johnson 2004, 81). Leslie Wayne claims that 'At the moment, the Pentagon has committed itself to more weapons systems than it can realistically afford, with the costs of many "megasystems" spiralling out of control. The Pentagon has about $1.3 trillion in weapon systems in some stage of development, with over $800 billion of those costs yet to be paid.' (Wayne 2005, C1) Other governments, which also have RMA aspirations, could find themselves in financially difficult situations. Whether the RMA at least affords them a military advantage remains to be seen.

Chapter 8

Conclusion

This book has tried to show the current degree of military privatization and to explain why private companies get increasingly involved in all aspects of warfare. It was further attempted to look at the implications of military privatization with respect to its military, economic and political dimensions and what it could all mean for international security. The main theme of the book was the influence of technology on the organization of the armed forces, their mode of operation and on security relationships more generally. It was argued that technology is a driving factor in the current process of military privatization because growing technological complexity exceeds the military's capability of managing it, especially in times of rapid technological change and times of diminishing internal resources. This current tendency of a privatization of military activities is not unprecedented in military history. For example, there are certainly many similarities between the present situation and the military revolution that occurred in the late Middle Ages (Avant 2001). However, the present situation is also unique in the sense that military privatization happens parallel to an overall process of economic globalization. There seems to be a complex relationship between technology, privatization and globalization. Technology and privatization are connected because privatization is a form of self-organization that is necessary for managing growing complexity caused by technology. By transferring numerous tasks to external organizations the military can reduce some of the enormous complexity it has acquired and it can therefore become more efficient and effective in managing itself.

Privatization and globalization are connected because private companies have great incentives in expanding their business and in entering new markets. Private defence firms simply follow this basic economic truth. The export of military equipment and services is still much more regulated than other business branches, but globalization by its very nature undermines possibilities of national control. Especially services are much more easily traded than hardware, as the trade of services is particularly difficult to regulate and to monitor. This could lead to a world in which military power becomes 'fungible' (Singer 2003, 171) and in which deterrence as a concept of security becomes obsolete or ineffective (Mandel 2002, 46-48). However, at the moment it is too soon to speculate of corporate armies and a political world beyond the nation state.

This study has put technology at the very centre of the development of a new privatized military service industry. The author can be accused of having fallen into the trap of technological determinism. However, this is only true to a certain extent and despite all explanatory weaknesses of technological determinism, it remains an 'intuitively compelling idea that technological innovation is a major driving force of contemporary history, if not the primary driving force' (Smith and Marx 1996,

XIV). It was not assumed that technology will automatically produce certain results or a certain end stage. Rather the assumption was that technology has a life of its own. Technology is at least partially autonomous and impacts in numerous ways on society and social organization. The introduction of new technology very often leads to unanticipated and unintended consequences. Technology tends to become ever more complex, making it more challenging to master it. Technology has also a strong tendency to form complex networks of interaction that make it so hard to get rid of successful new inventions. For example, once the Internet is established and widely used for all kinds of economic, scientific, military, educational and recreational purposes, it is almost impossible to abolish it again. As it becomes more widespread, more applications for it are found and the more it becomes ingrained in society. This tendency is also true for the use of technology by military organizations. Once the atom bomb is invented, it is impossible to un-invent it, or to pretend it never existed. Despite the fact that nuclear weapons have not been used in war since the Second World War, they impacted tremendously on military organizations, doctrine and warfare in general. The same applies to the information revolution: once network-centric systems are developed and fielded, they are bound to become more comprehensive and widespread. They already impact significantly on military organizations and security relationships.

The more complex a system becomes, the less it is influenced by its environment. Instead highly complex systems often attempt to shape their environment. Military procurement decisions do often not reflect military requirements, but rather a desire to build what is technically feasible and then find applications for it. It is often about solutions in search of problems. An illustrative example is missile defence. Originally it was meant to be a missile shield against a Soviet attack. As the Soviet Union disappeared a new purpose for the developed technologies had to be found. Similarly most of the technology that now is considered integral to the RMA was already developed in the 1970s and 1980s for the very specific purpose of countering the huge Soviet armoured forces that threatened Western Europe. A new purpose for the new technology had to be found and that was the RMA, which has become an end in itself. Critics of the RMA have very often pointed out that it is largely irrelevant to the most likely military challenges of the twenty first century (Gentry 2002, 88-103). Martin van Creveld goes as far as to claim that Western high-tech weapons are ready for the scrap heap.[1] This view completely underestimates the dynamic relationship between system and environment. As the military system is less shaped by its environment, it tries to shape its environment according to its own requirements. Modern armed forces are now having, or are in the process of acquiring, RMA enabled capabilities and they are going to find a purpose for these new capabilities.

---

1 Martin van Creveld argues: ' … As low-intensity conflict rises to the dominance, much of what has passed for strategy during the last two centuries will be proven useless. The shift from conventional war to low-intensity conflict will cause many of today's weapon systems, including specifically that are the most powerful and most advanced, to be assigned to the scrap heap.' (van Creveld 1991, 205).

This search for a purpose for the RMA was certainly to no small extent behind the American decision to invade Iraq. New weapons and tactics could be tested almost under laboratory conditions. The Iraq War demonstrated the superiority of US weapons and tactics and since then many of the most modern armed forces have been busy ordering new equipment with similar performance characteristics, clearly in anticipation of another Iraq-like contingency. As there is no real military requirement for developing and maintaining high-tech armed forces, the amount of resources that can be politically mobilized towards this end is limited. In theory it would be possible to again nationalize the defence industries and support capabilities of the armed forces and to reverse military privatization completely. In reality this would be impossible for any government to finance and would also be politically, as well as socially unfeasible in a democracy. The extraordinary high costs of military high-tech have led to the end of what can be called 'military socialism': the centralized control of military production and support capabilities. The resulting 'military capitalism' is more efficient, but puts heavy constraints on the use of military force.

This means Western armed forces will only go to war under circumstances in which the military, political and economic risks can be minimized (Shaw 2005, 72). Wars involving the most modern armed forces will most likely become 'virtual' in the sense that they will be waged with few combatants and result in relatively few casualties like the Kosovo air campaign (Ignatieff 2000, 4). On the other hand, the armed forces and their high-tech societies rapidly move far away from equilibrium towards criticality. They are highly vulnerable to terrorism and other forms of asymmetric warfare, which 'can be described as a less complex force' strategy for exploiting the vulnerabilities of a more complex opponent' (Elhefnawy 2004, 169). Complexity generally leads to instability and unpredictability of a system. Growing global complexity caused by interconnectedness and interdependence does not imply that the most advanced nations will never again go to war with each other. It rather makes the global system more unstable. David Rowe has pointed at the fact that the world economy was similarly globalized at the eve of the First World War, but globalization did not prevent the war and even contributed to its outbreak (Rowe 2005, 447). According to Rowe, globalization can lead to systemic instability because it can '(1) generate systemic insecurity as all major powers become less able to mobilize the countries' abundant resources for defense; (2) undermine the ability of powers to practice effective deterrence; (3) magnify the importance of alliances to a state's security while undermining the credibility of the promises on which the alliance rests; and (4) magnify the threat posed by states that are least constrained by deepening integration into the world economy' (Rowe 2005, 433).

Of course, Rowe does not suggest that another major war is inevitable, but rather that major war remains a possibility also in times of growing global interdependence. However, the militarism of the European powers at the beginning of the twentieth century did little in terms of averting war and was probably one of its causes. Should there ever be a major war again, numerical or even technological superiority might be completely irrelevant for the outcome. It is possible for the most technologically advanced and complex armed forces to manage successfully wars against a very inferior enemy, but if two high-tech armed forces (or coalitions) were to fight each other anything could happen. It would be like two armoured knights fighting on a tight rope above an abyss.

Bibliography

Adams, G., G. Ben-Ari, J. Logsdon and R. Williamson (2004), *Bridging the Gap: European C4ISR Capabilities and Transatlantic Interoperability* (Washington, DC: National Defense University).

Adams, J. (1998), *The Next World War: Computers Are the Weapons & the Front Line Is Everywhere* (New York: Simon & Schuster).

Adams, T.K. (1999), 'The New Mercenaries and the Privatization of Conflict', *Parameters US Army War College Quarterly* **29**(2): 103–116.

——, (2001), 'Future Warfare and the Decline of Human Decisionmaking', *Parameters US Army War College Quarterly* **31**(4): 57–71.

AgustaWestland (2004), 'Rotorsim Expands Its Capabilities With AB139 Flight Simulator', *AgustaWestland Press Release*, July 21, 2004.

Alberts, D.S. and T.J. Czwerwinski (eds) (1997), *Complexity, Global Politics, and National Security* (Washington, DC: National Defense University).

Allen, T.B. (1994), *War Games: Inside the Secret World and Men Who Play World War III* (London: Mandarin Paperbacks).

Ante, S.E. and S. Crock (2004), 'The Other U.S. Military; The Private Contractor Biz Is Hot, Vast, and Largely Unregulated. Is It Out of Control?', *Business Week* (3885), 31 May 2004: 76.

Armbruster, S. (2001), 'Private Sector's Military Bid', *BBC News Online*, 3 December 2001, available at <http://news.bbc.co.uk/1/hi/business/1689168.stm>, accessed 05/07/2006.

Arnold, G. (1998), *Mercenaries: Scourge of the Third World* (London: Palgrave Macmillan).

Arquilla, J. and D. F. Ronfeldt (1997), *In Athena's Camp: Preparing for Conflict in the Information Age* (Santa Monica: National Defense Research Institute Publication).

——. (2001), *Networks and Netwars: The Future of Terror, Crime, and Militancy* (Santa Monica: RAND).

Associated Press (2004a), 'Military Conflict Becoming A Private Enterprise', *The Associated Press*, 31 October 2003, available at <http://www.countercurrents.org/iraq-ap311003.htm>, accessed 05/07/2006.

——, (2004b), 'Elite U.S. Soldiers Leave for Better-Paying Jobs', *Associated Press*, 20 July 2004.

——, (2005), 'Will Navy's Sleek New Ship Help Fight the War on Terror?' *Associated Press*, 6 August 2005.

Avant, D. (2001), 'Selling Security: Post Cold War Private Security Services in Historical Perspective', San Francisco: Meeting of the American Political Science Association.

——, (2002), 'Privatizing Military Training', *Foreign Policy in Focus* **7**(6).
——, (2004), 'Think Again: Mercenaries', *Foreign Policy*, July–August 2004: 20–26.
——, (2004), 'The Outsourced War 'Is Here to Stay', *Business Week Online*, 21 May 2004.
——, (2005), *The Market for Force: The Consequences of Privatizing Security* (Cambridge: Cambridge University Press).
Babcock International (2003), *Annual Report*.
Bacevich, A.J. (2005), *The New American Militarism: How Americans Are Seduced by War* (Oxford: Oxford University Press).
Baker, J.C., K.M. O'Connell and J.A. Venzor (2002), 'Commercial Observation Satellites: Broadening the Sources of Geospatial Data', *Optical Engineering* **41**(9): 2077–2082.
Balaker, T. (2003), 'The New Art of War: How One Contractor Helps the World's Best Military Get Even Better', *Privatization Watch*, June/July 2003: 3 and 12.
Bamford, J. (1983), *The Puzzle Palace* (New York: Penguin).
Bateman, R.L. (ed.) (1999), *Digital War: A View From the Front Lines* (Novato, CA: Presidio Press).
Battilega, John, R. Greenwalt, D. Beachley, D. Beck, R. Driver and B. Jackson (2000), *Transformation in Defense Markets and Industries: Implications for the Future of Warfare*, National Intelligence Council (CIA) (published online 2000) <http://www.dni.gov/nic/research_defensemkts.html>, accessed 05/07/2006.
Baylis, J. and S. Smith (eds) (2001), *The Globalization of World Politics* (Oxford: Oxford University Press).
BBC (2000), 'Canada Seizes US Ship', *BBC News*, 4 August 2000, available at <http://news.bbc.co.uk/1/hi/world/americas/865091.stm>, accessed 05/07/2006.
——, (2001a), 'Privately Financing War', *BBC News Online*, 3 December 2001, available at <http://news.bbc.co.uk/1/hi/business/1666235.stm>, accessed 05/07/2006.
——, (2001b), 'Security Concerns Are Halted Behind the Frontline', *BBC World Service*, 3 December 2001, available at <http://news.bbc.co.uk/1/hi/business/1689031.stm>, accessed 10/11/2003.
——, (2001c), 'Private Company Bonanza', *BBC News Online*, 3 September 2001, available at <http://news.bbc.co.uk/1/hi/in_depth/business/2001/ppp/1504247.stm>, accessed 05/07/2006.
——, (2004), 'Missile Error Leads to RAF Deaths', *BBC Online*, April 9, 2004.
Beason, J.D. (1997), *DOD Science and Technology: Strategy for the Post-Cold War Era* (Washington, DC: National Defense University Press).
Ben-Ari, G. (2005), 'C3 Interoperability in Europe: The Challenge Ahead', *EuroFuture Magazine*, winter 2005.
Benbow, T. (2004), *The Magic Bullet: Understanding the Revolution in Military Affairs* (London: Brassey's).
Berkowitz, B.D. (2000), 'War Logs On – Girding America for Computer Combat', *Foreign Affairs* **79**(3): 8–11.
——, (2000), 'Information Warfare: Time to Prepare', *Issues in Science and Technology* **17**(2): 37–46.

——, (2003), *The New Face of War: How War Will Be Fought in the 21st Century* (New York: The Free Press).

Berkowitz, B.D. and A. Goodman (2000), *Best Truth: Intelligence in the Information Age* (New Haven: Yale University Press).

Bialos, J.P. and S.L. Koehl (2005), 'The NATO Response Force: Facilitating Coalition Warfare through Technology Transfer and Information Sharing', Washington, DC: *Center for Technology and National Security Policy/ National Defense University*, August 2005.

BICC (2004), *Conversion Survey 2004: Global Disarmament, Demilitarization and Demobilization* (Baden-Baden: Nomos Verlagsgesellschaft).

Biesecker, C. (2003), 'Demand For Contractor Services Increasing, CACI's London Says', *Defense Daily* **220**(10).

Bischak, G. (1999), 'Contending Security Doctrines and the Military Industrial Base', in A. Markusen and S.S. Castigan (eds), *Arming the Future: A Defense Industry for the 21st Century* (New York: Brookings University Press) 37–73.

Bitzinger, R.A. (2003), *Towards a Brave New Arms Industry?* (Oxford: Oxford University Press) Adelphi Paper (354).

Black, J. (2001), *War in the New Century* (New York/London: Continuum).

——, (2000), *War: Past, Present & Future* (Gloucestershire: Sutton Publishing Ltd.).

Bliss, G.L. (2002), 'The Revolution in Business Affairs: Strategic Business Alliances in Army Transformation', Carlisle Barracks, PA: *US Army War College*, Strategy Research Project.

Boeing (2004), *Annual Report*.

Boeing (2004a), 'Boeing Airborne Laser Team Achieves First Flight', *Press Release*, December 3, 2004.

Boese, W. (2005), 'Missile Defense Performance Top Concern', *Arms Control Today* **35**(4), May 2005: 31.

Bolkcom, C. and S. Squassoni (2002), 'Cruise Missile Proliferation: CRS Report to Congress', Washington D.C., 3 July 2002.

Bongard, D.L. and T.L. Sayers (2002), 'Impact of Robotics on Nontraditional Warfare', in W. Schilling (ed.), *Nontraditional Warfare: Twenty-First Century Threats and Responses* (Washington, DC: Brassey's) 299–309.

Book, E. (2002), 'Transportation Policy Chief Aiming for Higher Efficiency', *National Defense Magazine*, September 2002.

Bredemeier, K. (2003), 'Thousands of Private Contractors Support U.S. Forces in Persian Gulf-War', *Washington Post*, 3 March, 2003: E01.

Briody, D. (2003), 'Take a Look at the Carlyle Group: An Interview With Dan Briody', *Buzzflash*, 23 June 2003.

——, (2004), *The Halliburton Agenda: The Politics of Oil and Money* (Hoboken, NJ: John Wiley & Sons).

Broad, W.J. (2000), 'Antimissile Testing is Rigged to Hide a Flaw, Critics Say', *New York Times*, 9 June 2000.

Bromberg, J.L. (1999), *NASA and the Space Industry* (Baltimore: Johns Hopkins University Press).

Brown, S.H. (1999), 'Using Third Party Logisticians', *Army Logistician* **31**(6), November–December 1999.

Bundesrechnungshof (2003), 'Bemerkungen zur Haushalts- und Wirtschaftsführung des Bundes 2000', Berlin.

C4ISR Journal (2005), 'UK Lays Ground for Networked Future', <http://www.C4ISRJournal.com>, May 9, 2005, accessed 10/04/2006.

CAE (2005), *Military Simulation and Training*, Company Brochure, available at <http://www.cae.com/www2004/Products_and_Services/Military_Simulation_and_Training/Brochures/TTS_CAE_capability.pdf>, accessed 20/08/2005.

Cahlink, G. (2002), 'Army of Contractors', *Government Executive* **34**(2), February 2002.

——, (2005), 'Base Realignment and Closure', *GovExecLive*, April 13, 2005, interview available at <http://www.govexec.com/gelive/archive/20050413.htm>, accessed 15/09/05.

Caldicott, H. (2002), *The New Nuclear Danger: George W. Bush's Military-Industrial Complex* (New York: The New Press).

Callum, R. (1998), 'Dogfight: Exporting Supersonic Combat Aircraft in Post-Cold War Era', in G.I. Susman and S. O'Keefe (eds), *The Defense Industry in the Post-Cold War Era: Corporate Strategies and Public Policy Perspectives* (London: Pergamon: 103–120).

Camm, F. and V.A. Greenfield (2004), *How Should the Army Use Contractors On the Battlefield? Assessing Comparative Risk in Sourcing Decisions* (Santa Monica, CA: RAND Arroyo Center).

Campbell, D.W. (2002), 'The Privatization of Military Training Would Benefit the US', *National Defense Magazine*, November 2002.

Campbell, G. L. (2000), 'Contractors on the Battlefield: The Ethics of Paying Civilians to Enter Harm's Way and Requiring Soldiers to Depend upon Them', *Joint Services Conference on Professional Ethics*, Springfield, VA: January 27–28, 2000.

Canada House of Commons/S. Fraser (2001), 'National Defence – In-Service Equipment. Canadian Forces Cannot Guarantee Readiness of Combat Equipment', *Report of the Auditor General*, December 2001, available at <http://www.oag-bvg.gc.ca/domino/media.nsf/html/01pr09_e.html>, accessed 02/03/2006.

Canada DND (2004), 'Canadian Forces Contractor Augmentation Program', *DND website*, 14 July 2004, available at www.forces.gc.ca/site/Newsroom/view_news_e.asp?id=1409, accessed 04/07/2005.

Canadian Senate (2002), 'Managing and Marketing the Goose Bay, Labrador, Air Field', *Standing Senate Committee on National Finance*, July 2002, available at <http://www.parl.gc.ca/37/1/parlbus/commbus/senate/com-e/fina-e/rep-e/rep20jun02-e.htm>, accessed 02/03/2006.

Cappachio, T. (2003), 'Defense Contractors Benefit From War Spending', *Bloomberg News*, 2 April 2003, available at <http://seattlepi.nwsource.com/business/115323_raytheon02.shtml>, accessed 16/06/2006.

Cardinali, R. (2001), 'Does the Future of Military Logistics Lie in Outsourcing and Privatization? Accountants – the New Gatekeepers of War-Time Operations', *Work Study* **50**(3): 105–111.

Carter, A.B. and J.P. White (eds)(2000), *Keeping the Technological Edge: Managing Defense for the Future* (Cambridge, MA: The MIT Press).

Carter, A.B., M. Lettre and S. Smith (2001), 'Keeping the Technological Edge', in A.B. Carter and J.P. White (eds), *Keeping the Technological Edge: Managing Defense for the Future* (Cambridge, MA: The MIT Press) 129–163.

Carter, P. (2004), 'Hired Guns: What to Do About Military Contractors Run Amok', *Slate.msn.com*, 9 April 2004, available at <http://www.slate.com/id/2098571>, accessed 05/07/2006.

Castillo, L. A. (2000), 'Waging War with Civilians: Asking Unanswered Questions', *Aerospace Power Journal* **14**(3): 26–31.

Cebrowski, A.K. and J.W. Raymond (2005), 'Operationally Responsive Space: A New Defense Business Model', *Parameters US Army War College Quarterly* **34**(2): 67–77.

Chatterjee, P. (2004), *Iraq Inc.: A Profitable Occupation* (New York: Seven Stories Press).

Chesbrough, H.C. (2004), 'A Failing Grade for the Innovation Academy', *Financial Times*, 24 September 2004.

——, (2005), 'Breakthrough Ideas for 2005: Towards a New Services Science', *Harvard Business Review* **83**(2): 43–44.

Chin, W. (2001), 'The Impact of Communications and the Media on the Art of War since 1815', in G. Jensen and A. A. Wiest (2001), *War in the Age of Technology: Myriad Faces of Modern Armed Conflict* (New York: New York University Press).

Chisholm, P. (2002), 'A Vast Improvement in Navy Training at Sea', *Military Training Technology Online Edition* **7**(8), December 2002.

Chu, D.S. and M.C. Waxman (1998), 'Shaping the Structure of the American Defense Industry', in G.I. Susman and S, O'Keefe (eds), *The Defense Industry in the Post-Cold War Era: Corporate Strategies and Public Policy Perspectives* (London: Pergamon) 35–44.

Cioffi-Revilla, C. (2000), 'War and Warfare: Scales of Conflict in Long-Range Analysis', in R.A. Denemark, J. Friedman, B.K. Gills and G. Modelski (eds), *World System History/ The Social Science of Long-Term Change* (London: Routledge).

Clark, E. (2001), 'Private Finance on an International Scale', *BBC News Online*, 31 August 2001.

Clark, W.K. (2003), *Winning Modern Wars: Iraq, Terrorism and the American Empire* (New York: Public Affairs).

Cohen, E.A. (1996), 'A Revolution in Warfare', *Foreign Affairs* **75**(2): 37–54.

——, (2002), 'Technology and Warfare', in J. Baylis (ed.), *Strategy in the Contemporary World: An Introduction to Strategic Studies* (Oxford: Oxford University Press) 235–253.

Cohen, N. (2003), 'The Mercenaries Who Fight for Britain', *New Statesman*, 27 January 2003: 20–21.

Coker, C. (2001), *Humane Warfare* (London: New York, Routledge).

——, (2002), *Waging War Without Warriors? The Changing Culture of Military Conflict* (Boulder, CO: Lynne Rienner Publishers).

——, (2004), *The Future of War: The Re-Enchantment of War in the Twenty-First Century* (London: Blackwell).

Collier, C.A. (2001), 'A New Way to Wage War: US Support to Operation *Stabilise'*, *Military Review*, January–February 2001.

Context (2003), 'Defense Exports: The JSF Ideology', *Context* No. 58, January 2003, available at <http://www.dedefensa.org/article.php?art_id=677>, accessed 07/07/2006.

Cook, M. L. (2004), *The Moral Warrior: Ethics and Service in the U.S. Military* (New York: State University of New York Press).

Cordesman, A.H. (2003), *The Iraq War: Strategy, Tactics, and Military Lessons* (Westport, CT: Praeger).

Correll, J.T. and C.T. Nash (1991), 'The Industrial Base at War', *AirForce Magazine* **74**(12), December 1991.

Correll, J. T. (1998), 'Long Range Blind Spot: There Is No Plan for Long Range Airpower Beyond Upgrades and Improvements for the Existing Bomber Fleet', *Air Force Association/ Editorial* **80**.

Cray, C. and L. Drutman (2005), 'Corporations and the Public Purpose: Restoring the Balance', *Seattle Journal of Social Justice* **4**(1), fall/ winter 2005.

Crock, S. (2003), 'Five Truths about Boeing's Tanker Lease', *Business Week*, 30 June 2003.

Cuadrado-Roura, J.R., L. Rubalcaba-Bermejo and J.R. Bryson (eds) (2002), *Trading Services in the Global Economy* (Celtenham: Edward Elgar Publishing Ltd.).

Cullather, N. (2003), 'Bombing at the Speed of Thought: Intelligence in the Coming Age of Cyberwar', *Intelligence and National Security* **18**(4): 141–154.

Davis, J. R. (2000), *Fortune's Warriors: Private Armies and the New World Order* (Vancouver: Douglas McIntyre).

Davis, J. (2003), 'If We Run Out of Batteries, This War Is Screwed', *Wired*, Issue 11.06, June 2003.

Defence Systems Daily (2001), 'MoD Lets Astute Class Training Contract', *Defence Systems Daily*, 13 September 2001, available at <http://defence-data.com/dsei/pageds1032.htm>, accessed 29/06/2004.

Defense & Aerospace Electronics, 'ADS Offers Great Opportunity: DMSO Chief', *Defense & Aerospace Electronics*, 30 August 1993.

Defense Daily (2002), 'CSC Selected For $6.1 Billion German Army IT Project', *Defense Daily*, 7 June 2002.

——, (2006), '24-Year, GBP 1.5B Contract to Maintain UK's Hercules Fleet', *Defense Industry Daily*, June 6, 2006, available www.defenseindustrydaily.com/2006/06/24year-gbp-15b-contract-to-maintain-uks-hercules-fleet/index.php, accessed 06/06/2006.

Defense Industry Daily (2006), 'EADS Wins Eur 175M Contract for French Air Force Training Support', *Defense Industry Daily*, 27 April 2006.

Defense Week (2002), 'CAE Wins U.S. Coast Guard, German Forces Services Work', *Defense Week* **23**(13).

DeGroot, G. (2004), *The Bomb: A Life* (London: Jonathan Cape).

DeLanda, M. (1991), *War in the Age of Intelligent Machines* (New York: Zone Books).

Demchak, C.C. (1991), *Military Organizations, Complex Machines: Modernization in the U.S. Armed Forces* (New York: Cornell University Press).

——, (1992), 'Complexity, Rogue Outcomes and Weapon Systems', *Public Administration Review*, July/August 1992: 347–356.

——, (1999), '"New Security" in Cyberspace: Emerging Intersection between Military and Civilian Contingencies', *Journal of Contingencies and Crisis Management* **7**(4).

Demchak, C.C. and S.E. Goodman (1995), 'Computers as Substitute Soldiers?' *Communications of the ACM* **38**(2), February 1995.

Demchak, C.C. and G.I. Rochlin (1991), *Lessons of the Gulf War: Ascendant Technology and Declining Capability* (Berkeley: University of California. Policy Papers in International Affairs).

Denning, D.E. (2003), 'Cyber-Security as an Emergent Infrastructure', in R. Latham (ed.), *Bombs and Bandwith: The Emerging Relationship Between Information Technology and Security* (New York: The New Press).

DerDerian, J. (2000), 'War Games: The Pentagon Wants What Hollywood's Got', *The Nation*. **270**(13): 40–43.

——, (2001), *Virtuous War: Mapping the Military-Industrial-Media-Entertainment Network* (Boulder, CO: Westview Press).

Dewitte, L. (2006), 'Turkey Offers Indonesia F-16 Jetfighter Maintenance', *F-16.net*, April 24, 2006, available at <www.f-16.net/news_article1765.html>, accessed 24/06/2006.

Dippold, M.K. (1997), 'Air Occupation: Asking the Right Questions', *AirPower Journal*, winter 1997.

Divis, D.A. (2002), 'Military Role Emerges for Galileo', *GPS World*, May 1, 2002, available at <http://www.gpsworld.com/gpsworld/article/articleDetail.jsp?id=21977>, accessed 14.03.2006.

Doborek, C.J. (2002), 'Intercepts: Games Militaries Play', *Federal Computer Week*, 26 August 2002.

Dolman, E.C. (2005), *Pure Strategy: Power and Principle in the Space and Information Age* (London: Frank Cass).

Dombrowski, P.J., E. Gholz and A.L. Ross (2003), *Military Transformation and the Defense Industry After Next the Defense Industrial Implications of Network-Centric Warfare* (Newport: Naval War College Press).

Donnelly, H. (2004), 'ACS Fight Nears Decision', *Military Information Technology* **8**(2), 15 March 2004.

Douglas, K.L. (2004), 'Contractors Accompanying the Force: Empowering Commanders with Emergency Change Authority', *Air Force Law Review*, spring 2004 (55): 127–156.

Dowd, A.W. (2003), 'Thinking the Unthinkable, 2.0', *Hudson Institute*, 19 December 2003.

Downs, A. (1957), *An Economic Theory of Democracy* (New York: Harper).

Drake, J. (1999), 'Fleet Management: The Mother of All Contracts: Army to Relinquish M109 Howitzer, Apache Day-to-Day Care to Industry in Pilot Program', *National Defense Magazine*, January 1999.

Drohan, M. (2004), Making a Killing: Why and How Corporations Use Armed Forces to Do Business (The Lyons Press).

Dunnigan, J. (2006), 'Get the T-90s Out of the Kitchen', *StrategyPage*, June 24, 2006, available at <http://www.strategypage.com/dls/articles/2006624232423.asp>, accessed 06/08/2006.

Dupre, J.S. and W.E. Gustafson (1962), 'Contracting for Defense: Private Firms and the Public Interest', *Political Science Quarterly* **77**(2), June 1962.

Dye, P.J. (2001), 'Maintenance Support Strategies', *Air Force Journal of Logistics*, spring 2001.

EADS (2004a), 'SKYNET 5 PFI Wins Accolades', *EADS Press Release*, February 2, 2004.

EADS (2004b), 'SATCOMBw: Satellite Communications for the German Armed Forces', *EADS Press Release*, 10 May 2004.

Ebbutt, G. (ed.) (2004), *Jane's C4I Systems 2004-2005* (Coulsdon: Jane's Information Group).

Economist (2003), 'Military-Industrial Complexities; Defence Contractors. (The Defence Industry and the War)', *The Economist (US)* **366**(8317), 27 March 2003.

——, (2004), 'Dangerous Work: Private Security Firms in Iraq', *The Economist*, 7 April 2004.

——, (2005a), 'Under New Management: America's Atomic Bomb Laboratories Are About to Be Overhauled. It Promises to Be a Tricky Business', *The Economist* **8434**, 7 July 2005: 71.

——, (2005b), 'A Good Enough Performance? America's Most Famous Weapons Laboratory Is Under New Management', *Economist* **379** (8482): 85–86.

——, (2006), 'The March of Technology: Military Equipment: Researchers Are Devising New Ways to Turn Soldiers into Cyborg-like, High-tech Fighting Machines', *The Economist* **379**(8481), 10 June 2006: 27–28.

Edmonds, B. (1999), 'What is Complexity? - The Philosophy of Complexity *Per Se* with Application to Some Examples in Evolution', in F. Heylighen and D. Aerts (eds), *The Evolution of Complexity* (Kluwer: Dordrecht).

Ehab, M.S., D.M. Cox, L.L. Delorme and C.Y. Lanclos (1999), 'Logisticians Don't Always Wear Green – Contractors as Force Multipliers', *Quartermaster Professional Bulletin*, summer 1999.

Eisendrath, C., M.A. Goodman and G.E. Marsh (2001), *The Phantom Defense: America's Pursuit of the Star Wars Illusion* (Westport, CT: Praeger).

Elbert, B. (2005), 'Competition and Consolidation in the Satellite Launch Sector', *SatMagazine.com*, June 2005.

Elhefnawy, N. (2004), 'Societal Complexity and Diminishing Returns in Security', *International Security* **29**(1): 152–174.

——, (2003), 'Four Myths about Space Power', *Parameters US Army War College Quarterly* **33**(1): 124–132.

Eunjung Cha, A. and R. Merle (2004), 'Line Increasingly Blurred Between Soldiers and Civilian Contractors', *Washington Post*, 13 May 2004: A01.

European Union (2005), 'Defence Procurement in the European Union/ The Current Debate', *European Union Institute for Security Studies*, Report of an EUISS Task Force, Chairman and Rapporteur Burkard Schmitt, May 2005.

Evans, M., R. Parkin and A. Ryan (eds) (2004), *Future Armies, Future Challenges: Land Warfare in the Information Age* (Crows Nest, NSW: Allen Unwin).

Evans, M. (2004), 'Australia's Approach to the Revolution in Military Affairs, 1994–2004', in E.O. Goldman and T.G. Mahnken (eds), *The Information Revolution in Military Affairs in Asia* (New York: Palgrave Macmillan: 23–56).

Faite, A. (2004), 'Involvement of Private Contractors in Armed Conflict: Implications under International Humanitarian Law', *Defence Studies* **4**(2): 166–183.

Farrell, T. (1997), *Weapons Without a Cause: The Politics of Weapons Acquisition in the United States* (London: Macmillan Press).

Fallows, J. (2002), 'Uncle Sam Buys an Aircraft', *The Atlantic Monthly* **289**(6): 62–74.

Felkins, S. L. (2001), 'Outsourced Tyranny: The Use of Private Companies to Do Government's Dirty Work', *Antiwar.Com*, available at <http://antiwar.com/orig/felkins2.html>, accessed 02/03/2006.

Ferris, J. (2003), 'A New American Way of War? C4ISR, Intelligence and Information Operations in Operation 'Iraqi Freedom': A Provisional Assessment', *Intelligence and National Security* **18**(4): 155–174.

Ferris, S.P. and D.M. Keithly (2001), 'Outsourcing the Sinews of War: Contractor Logistics', *Military Review* **81**(5): 72–83.

Fisher, C. (2005), 'Simulations Add "Playtime" to Training Cycle: Simulations Take Some of the "Field" Out of Training', *Marines Magazine*, Jan–March 2005: 16–24.

Fitzpatrick, A. (1998), 'Teller's Technical Nemesis: The American Hydrogen Bomb and its Development Within a Technological Infrastructure', *Phil & Tech* **3**(3): 10–17.

Forster, A. and W. Wallace (2001), 'What Is NATO For?' *Survival* **43**(4): 107–122.

Fortner, J. A. (2000), 'Institutionalizing Contractor Support on the Battlefield', *Army Logistician* **32**(4): 12–15.

Fox, B., M. Boito and O. Younossi (2004), 'Test and Evaluation Trends and Costs for Aircraft and Guided Weapons', Santa Monica, CA: *RAND*.

France, M.E.B. (2000), 'Back to the Future: Space Power Theory and A.T. Mahan', *Air & Power Chronicles*, August 4, 2000, available at <http://www.airpower.maxwell.af.mil/airchronicles/cc/france1.html>, accessed 16/03/2006.

Freedman, L. (1998), *The Revolution in Strategic Affairs* (Oxford: Oxford University Press, Adelphi Paper (318)).

Freedberg, S.J. (2001), 'Pentagon War Games Grow More Complex', *GovExec.Com*, 11 August 2001.

——, (2003), 'Army's Lightweight Stryker Brigade About to Be Tested in Iraq', *GovExec.Com*, 10 October 2003.

Friedman, R. M. (2002), *Civilian Contractors on the Battlefield: A Partnership With Commercial Industry or Recipe for Failure?* (Carlisle Barracks, PA: U.S. Army War College).

Friedman, G. and M. Friedman (1996), *The Future of War: Power, Technology, and American World Dominance in the 21st Century* (New York: Crown Publishers).

Fukuyama, F. (1992), *The End of History and the Last Man* (New York: Macmillan).

Fuller, J.F.C. (1998), *Armament & History: The Influence of Armament on History From the Dawn of Classical Warfare to the End of the Second World War* (New York: Da Capo Press).

Galace, P.I. (2005), 'Asian Satellite Companies: Waiting for the Cavalry to Charge', *SatMagazine.com*, June 2005.

Galbraith, J.K. (1999), 'Free Market Fraud', *The Progressive*, January 1999, available at <http://www.progressive.org/mag_galbraith0199>, accessed 05/04/2005.

Gallagher, P. (2005), 'Sellafield Plutonium 'Missing' ', *Mirror*, 18 February 2005.

Gansler, J. S. (1998), 'The Revolution in Business Affairs: the Need to Act Now', *Program Manager*, September–October: 30–33.

Gegax, T.T. (2004), 'War Stories: Military Moonlighting', *Newsweek Web Exclusive*, 8 June 2004.

Gentry, J.A. (2002), 'Doomed to Fail: America's Blind Faith in Military Technology', *Parameters US Army War College Quarterly* **32**(4): 88–103.

German Ministry of Defence (BMVg)/Bundeswehr (2005), 'Heeresinstandsetzungslogistik', *Press Release*, 27 January 2005.

Giddens, A. (1990), *The Consequences of Modernity* (Stanford: Stanford University Press).

Ghamari-Tabrizi, S. (2000), 'Simulating the Unthinkable: Gaming Future War in the 1950s and 1960s', *Social Studies of Science* **30**(2): 163–224.

——_, (2005), *The Worlds of Herman Kahn: The Intuitive Science of Thermonuclear War* (Cambridge, MA: Harvard University Press).

Gholz, E. (2000), 'The Curtiss-Wright Corporation and Cold War-Era Defense Procurement: A Challenge to Military-Industrial Complex Theory', *Journal of Cold War Studies* **2**(1): 35–75.

——, (2002), 'America's Innovative Advantage: Systems Integration in the U.S. Defense Industry', *Meeting of the American Political Science Association, Boston*, August 2002.

Glantz, A. (2003), 'Private Company Manages Daily Bombing of Korean Village', *CorpWatch*, September 27, 2003, available at <http://www.corpwatch.org/article.php?id=8656>, accessed 05/07/2006.

Gluck, F. (2002), 'Candid Voices: Military Logistics and the Warfighter', *Air Force Journal of Logistics*, spring 2002.

Goldman, E.O. and T.G. Mahnken (eds)(2004), *The Information Revolution in Military Affairs in Asia* (New York: Palgrave Macmillan).

Gongora, T. and H. von Riekhoff (2000), *Toward a Revolution in Military Affairs? Defense and Security at the Dawn of the Twenty-First Century* (Westport, CT: Greenwood Press).

Government Executive (2005), 'Daily Briefing/ Pentagon Proposes to Close 33 Major Bases', *GovExec.com*, 13 May 2005.

Graham, B. (2003), *Hit to Kill: The New Battle Over Shielding America From Missile Attack* (New York: Public Affairs).

Gray, C.H. (1997), *Postmodern War: The New Politics of Conflict* (London: Routledge).

——, (2005), *Peace, War, and Computers* (London: Routledge).

Gray, C.S. (2003), *Strategy for Chaos: Revolutions in Military Affairs and The Evidence of History* (London: Frank Cass).

——, (2005), *Another Bloody Century: Future Warfare* (London: Weidenfeld & Nicolson).

Gregory, W. H. (1993), *The Price of Peace: The Future of Defense Industry and High Technology in a Post-Cold War World* (New York: Lexington Books).

Grimes, B. (2004), 'Defense Gaming Fine-tunes Soldiers' Skills', *Government Computer News* **23**(27): 36.

Guillory, M.E. (2001), 'Civilianizing the Force: Is the United States Crossing the Rubicon? – Role of Civilians Under the Laws of Armed Conflict', *Air Force Law Review* **51**(2001): 115–116.

Guma, G. (2004), 'Outside View: Privatizing War', *UPI*, 7 July 2003.

——, (2004), 'Outsourcing Defense: The Quiet Rise of the National Security, Inc.', *Towards Freedom Online*, summer 2004, available at <http://www.towardfreedom.com/home/content/view/449/69>, accessed 07/07/2006.

Hackemer, K. (2001), *The U.S. Navy and the Origins of the Military-Industrial Complex, 1847-1883* (Annapolis, MD: Naval Institute Press).

Hackworth, D. (2001), 'When Cannons Shoot, Call Brown & Root', *WorldNetDaily*, Commentary, 14 August 2001.

Haglund, D.G. and S.N. MacFarlane (ed.) (1999), *Security, Strategy and the Global Economics of Defence Production* (London: McGill-Queen's University Press).

Hall, R.B. and T.J. Biersteker (eds)(2002), *The Emergence of Private Authority in Global Governance* (Cambridge: Cambridge University Press).

Halliburton (2004), 'KBR Awarded Contract For CONLOG', *Press Release*, 3 February 2004.

Hallinan, C. (2003), 'Rumsfeld's New Model Army', *Foreign Policy in Focus*.

Hallion, R.P. (1992), *Storm Over Iraq: Air Power and the Gulf War* (Washington, DC: Smithsonian Institution Press).

Hambling, D. (2005), *Weapons Grade: The Revealing History of the Link Between Modern Warfare and Our High-Tech World* (London: Constable & Robinson Ltd.).

Hambleton, K. (2005), *Conquering Complexity: Lessons for Defence Systems Acquisition* (London: University College London, The Defence Engineering Group, published by the Stationary Office. Norwich).

Hammes, T.X. (2004), *The Sling and the Stone: On War in the 21st Century* (St. Paul, MN: Zenith Press).

Hammond, G. T. (1996), *Paths to Extinction: The US Air Force 2025* (Washington, DC: Department of the US Air Force).

Hari, J. (2003), 'This Nightmare World Full of Privatised Armies', *The Independent*, 14 November 2003: 21.

Harkavy, R.E. and S.G. Neuman (2001), *Warfare and the Third World* (New York: Palgrave).

Harris Corp. (2003), 'Harris Corporation Selected As Preferred Supplier to the Canadian Forces Communications and Information Services', *Press Release*, 31 March 2003.

——, (2004), 'U.S. National Reconnaissance Office Selects Harris Corp. for $1 Billion Communications Operations and Maintenance Program', *Press Release*, 26 August 2004.

Harris, M. (2000), *LOGCAP: The Nation's Premier Contingency Contracting Program for Force XXI* (Carlisle Barracks, PA: U.S. Army War College).

Harris, P. (1996), 'Military Advising Is Growth Industry', *Insight*, August 26, 1996: 12–14.

Harris, S. (2004), 'Playing for Keeps', *GovExec.Com*, 1 August 2004.

——, (2006), 'Contractor's Rise Shows Blurred Government, Industry Lines', *National Journal*, 7 July 2006, available at <http://govexec.com/dailyfed/0706/070706nj1.htm>, accessed 09/07/2006.

Hartley, K. (2004), 'The Economics of Military Outsourcing', *Defence Studies* **4**(2): 199–206.

Hartung, W.D. (2003), *How Much Are You Making on the War, Daddy? A Quick and Dirty Guide to War Profiteering in the Bush Administration* (New York: Nation Books).

——, (2004), 'Private Military Contractors in Iraq and Beyond: A Question of Balance', *World Policy Institute*, available at <http://www.worldpolicy.org/projects/arms/updates/FPIFJune2004.html>, accessed 15/03/2006.

——, (2006), 'Contractors versus Soldiers: Emerging Budgetary Reality', *World Policy Institute*, Report, 10 February 2006, available at <http://www.worldpolicy.org/projects/arms/reports/soldiers.html>, accessed 15/03/2006.

Harz, C. (2005), 'Network Centric Warfare: The European Conference', available at <http://www.iqpc.com/binary-data/ IQPC_CONFEVENT/pdf_file/9787.doc>, accessed 22/06/2006.

Hau, D. Y. (2003), 'Conscription and Force Transformation', *Pointer: Singapore Armed Forces Journal* **29**(4), available at <http://www.mindef.gov.sg/imindef/publications/pointer/journals/2003/v29n4/features/feature2.html>, accessed 07/07/2006.

Hayward, K. (2000), 'The Globalisation of Defence Industries', *Survival* **42**(2), summer 2000: 115–132.

——, (2005), '"I Have Seen the Future and It Works": The US Defence Industry Transformation – Lessons for the UK Defence Industrial Base', *Defence and Peace Economics* **16**(2): 127–141.

Held, D., A. McGrew, D. Goldblatt and J. Perraton (1999), *Global Transformations: Politics, Economics, and Culture* (Cambridge: Polity Press).

Heinrich, T. (2002), 'Cold War Armory: Military Contracting in Silicon Valley', *Enterprise & Society* **3**, June 2002.

Herman, M. (1998), 'Entropy Based Warfare: Modeling the Revolution in Military Affairs', *Joint Forces Quarterly*, autumn/winter 1998/99: 85–90.

Hess, P. (2005), 'US: Pentagon to Privatize Security of Military Bases in Europe', *UPI*, February 18, 2005.

Hewson, R. (2005), 'Cruise Missile Technology Proliferation Takes Off', *Jane's Intelligence Review* **17**(10): 41–45.

Hills, A. (2004), *Future War in the Cities: Rethinking a Liberal Dilemma* (London: Frank Cass Publishers).

Hirsh, M. and J. Barry (2005), 'Nukes: Nuclear Offense', *NEWSWEEK*, 21 February 2005: 16–20.

Howard, M. (2001), *War in European History* (Oxford: Oxford Paperbacks).

Huber, P. (2003), 'Military-Industrial Complex, 2003', *Forbes.Com*.

Huntington, S.P. (1957), *The Soldier and the State* (Cambridge, MA: Press of Harvard).

Hutchinson, R. (2003), *Weapons of Mass Destruction: the Non-Nonsense Guide to Nuclear, Chemical and Biological Weapons Today* (London: Weidenfeld & Nicholson).

Ignatieff, M. (2000), *Virtual War: Kosovo and Beyond* (London: Chatto & Windus).

Irwin, S.I. (2000), 'Pentagon's Computer Security Dilemmas Remain Unsolved', *National Defense Magazine*, July 2000.

——, (2001), 'Forecast is Rosy for Visual Simulation Industry', *National Defense Magazine*, November 2001.

——, (2003), 'Army Logistics: Changes Ahead in Force Structure, Maintenance', *National Defense Magazine*, January 2003.

Isenberg, D. (1997), 'Soldiers of Fortune Ltd.: Profile of Today's Private Sector Corporate Mercenary Firms', *Center for Defense Information*, Washington, DC.

——, (2004), 'Corporate Mercenaries in Iraq/ Part 1: Profit Comes at a Price', *Asia Times Online*, 19 May 2004.

Ismail, M.A. (2004), 'The Sincerest Form of Flattery: Private Equity Firms Follow in Carlyle Group's Footsteps', *Center for Public Integrity*, 18 November 2004, available at <http://www.publicintegrity.org/pns/report.aspx?aid=425>, accessed 15/05/2006.

James, A.D., D. Cox and J. Rigby (2005), 'Testing the Boundaries of Public Private Partnership: The Privatisation of the UK Defence Evaluation and Research Agency', *Science and Public Policy* **32**(2): 155–161.

Jane's Information Group (2005), 'No Buyers for UK's Defence Aviation Repair Agency (DARA) and Government Now Recommends Quick Closure', *Press Release*, April 8, 2005.

Jensen, G. and A. A. Wiest (2001), *War in the Age of Technology: Myriad Faces of Modern Armed Conflict* (New York: New York University Press).

Jervis, R. (1997), 'Complex Systems: The Role of Interactions', in D.S. Alberts and T.J. Czwerwinski (eds), *Complexity, Global Politics, and National Security* (Washington, DC: National Defense University).

Ji, Y. (2004), 'Learning and Catching Up: China's Revolution in Military Affairs Initiative', in E.O. Goldman and T.G. Mahnken (eds) (2004), *The Information Revolution in Military Affairs in Asia* (New York: Palgrave Macmillan) 97–117.

Johnson, C. (2004), *The Sorrows of Empire/ Militarism, Secrecy, and the End of the Republic* (London: Verso).

——, (2004b), 'The Arithmetic of America's Military Bases Abroad: To What Does It Add Up?', *History News Network*, January 19, 2004, available at <http://hnn.us/articles/3097.html>, accessed 15/06/2005.

Jones, W. (2001), 'The Value of Military Industrial Offsets', *DISAM Journal*, winter 2001–2002: 108.

Joyce, L. (2005), 'Military Apps Drive Simulation Tools: Cutting-Edge Software Gives Soldiers the Training They Need Before Heading to the Real "Playing Field"', *R & D* **47**(6): 16–19.

Kaldor, M. (1979), 'Defence Costs and the Defence Industry', in M. Kaldor, D. Smith and S. Vines (eds), *Democratic Socialism and the Costs of Defence: The Report and Papers of the Labour Party Defence Study Group* (London: 286).

——, (1983), *The Baroque Arsenal* (London: Abacus).

——, (1986), 'The Weapons Succession Process', *World Politics* **38**(4).

——, (1999), *New and Old Wars: Organised Violence in a Global Era* (Stanford: Stanford University Press).

——, (2000), *Global Insecurity: Restructuring the Global Military Sector III* (London/ New York: Pinter).

Kaldor, M., G. Schméder and U. Albrecht (1998), *Restructuring the Global Military Sector II: The End of Military Fordism* (London: Pinter).

Keefe, P.R. (2005), *Chatter: Dispatches from the Secret World of Global Eavesdropping* (New York: Random House).

Keegan, J. (1987), *The Mask of Command: A Study of Generalship* (London: Pimlico).

——, (1993), *A History of Warfare* (New York: Random House).

——, (1998), *War and Our World: The Reith Lectures* (London: Hutchinson).

Keller, W.W. (2001), 'Mortgaging Security for Economic Gain: U.S. Arms Policy in an Insecure World', *International Studies Perspectives*, **2001**(2): 177–193.

Kelley, M.R. and T.A. Watkins (1998), 'Are Defense and Non-Defense Manufacturing Practices So Different?' in G.I. Susman and S. O'Keefe (eds), *The Defense Industry in the Post-Cold War Era: Corporate Strategies and Public Policy Perspectives* (London: Pergamon) 251–280.

Kennedy, H. (2004), 'Forces Under Stress', *National Defense Magazine*, October 2004.

Kennedy, P. (1984), 'The First World War and the International Power System', *International Security* **9**(1): 7–40.

Keohane, R.O. and J.S. Nye (1998), 'Power and Interdependence in the Information Age', *Foreign Affairs* **77**(5): 81–94.

Kerber, R. (2004), 'For Army, a Clash Over Supply, Demand: Contractors Stymied Amid Shortage of Armoured Vehicles', *The Boston Globe*, 30 December 2004.

Kimball, C.G. (2005), 'Replacement Nuclear Warheads? Buyer Beware!' *Arms Control Today* **35**(4), May 2005, available at <http://www.armscontrol.org/act/2005_05/focus.asp>, accessed 07/07/2006.

Klare, M.T. (2002), *Resource Wars: The New Landscape of Global Conflict* (New York: Owl Books).

Klaus, J.D. (2004), 'CRS Report for Congress/ U.S. Overseas Basing: Background and Oversight Issues for Congress', *US Congressional Research Service*, 17 November 2004.

Knox, M. and W. Murray (2001), *The Dynamics of Military Revolution, 1300-2050* (Cambridge: Cambridge University Press).

Koenig, J. (2005), 'SALIS – Ein Multinationales Projekt zur Lösung der Militärischen Transporterfrage?', JetJournal.net, March 24, 2005, available at www.jetjournal.net/content/view/3397/35/, accessed 20/05/2006.

Korb, L. and C. Wadhams (2005), 'U.S. Should Set Example on Limiting Arms Exports', *Boston Globe*, 27 March 2005.

Krause, K. (1992), *Arms and the State: Patterns of Military Production and Trade* (Cambridge: Cambridge University Press).

Kreisher, O. (1999), 'The Revolution in Business Affairs: A Sea Change in Defense Procurement', *Seapower*: 71–77.

Krepinevich, A.F. (2002), *The Military-Technical Revolution: A Preliminary Assessment* (Washington, DC: Center for Strategic Budgetary Assessments, US DoD, Office of Net Assessment).

Kurlantzick, J. (2003), 'Outsourcing the Dirty Work', *The American Prospect* **14**(5), 1 May 2003: 17–19.

Kutner, J.A. (2001), 'Robust Weapon Simulations Hinge On Close Collaboration', *National Defense*, January 2001.

Laird, Robbin F. (2003), 'Transformation and the Defense Industrial Base: A New Model', *Defense Horizons*, No. 26, May 2003: 1–8.

Lang, P. (2006), 'The Vulnerable Line of Supply in Iraq', *Christian Science Monitor*, July 21, 2006, available at <http://www.csmonitor.com/2006/0721/p09s01-coop.html>, accessed 04/08/2006.

Lanning, M.L. (2005), *Mercenaries: Soldiers of Fortune, From Ancient Greece to Today's Private Military Companies* (New York: Presidio Press).

Latham, A. (2000), 'The Military-Industrial Correlates of the RMA: The Evolution of Agile Manufacturing', in T. Gongorra and H. von Riekhoff (eds), *Toward a Revolution in Military Affairs? Defense and Security at the Dawn of the Twenty-First Century* (London: Greenwood Press) 159–181.

Latham, N. (2003), *Defence Industry in a Global Context: Policy Implications for the United Kingdom* (London: The Royal United Services Institute).

Latham, R. (ed.) (2003), *Bombs and Bandwidth: The Emerging Relationship Between Information Technology and Security* (New York: The New Press).

Lavallee, T. (2003), 'Globalizing the Iron Triangle: Policy-Making Within the US Defense Industrial Sector', *Defense & Security Analysis* **19**(2): 149–164.

Lavelle, M., J. Rae-Dupree, P. Sherrid and C.H. Schmitt (2003), 'Battle-Ready Business. (Private Businesses' Roles in War on Iraq', *U.S. News & World Report*, 31 March 2003: 52–53.

Lenoir, T. (2000), 'All but War Is Simulation: The Military-Entertainment Complex', *Configurations* **8**(3): 289–336.

——, (2003), 'Programming Theaters of War: Gamemakers as Soldiers', in R. Latham (ed.), *Bombs and Bandwidth/ The Emerging Relationship Between Information Technology and Security* (New York: The New Press) 175–198.

Leonhard, R.L. (1998), *The Principles of War for the Information Age* (New York: Ballantine Books).

Libicki, M. C. (1994), *The Mesh and the Net: Speculations on Armed Conflict in a Time of Free Silicon* (Washington, DC: National Defense University).

——, (2000), 'What Is Information Warfare?' in T. Gongorra and H. von Riekhoff (eds), *Toward a Revolution in Military Affairs? Defense and Security at the Dawn of the Twenty-First Century* (London: Greenwood Press) 37–60.

Lippitz, M.J., S. O'Keefe and J.P. White (2000), 'Advancing the Revolution in Business Affairs', in A.B. Carter and J.P. White (eds), *Keeping the Technological Edge: Managing Defense for the Future* (Cambridge, MA: The MIT Press) 165–202.

Lockheed Martin (2002), 'Lockheed Martin and the British Army Unveil the World's Most Sophisticated Simulation System', *Press Release*, 27 November 2002.

Lockheed Martin (2004), *Annual Report.*

Lockheed Martin (2004a), 'Operation Iraqi Freedom', *Public Relations Material.*

Lonsdale, D. (2004), *The Nature of War in the Information Age: Clausewitzian Future* (New York: Frank Cass Publishers).

Lorell, M.A., J.F. Lowell, R.M. Moore, V.A. Greenfield and K. Vlachos (2002), *Going Global? U.S. Government Policy and Defense Aerospace Industry* (Santa Monica: RAND).

Lovering, J. (2000), 'Loose Cannons: Creating the Arms Industry', in M. Kaldor (ed.) (2000), *Restructuring the Global Military Sector/ Volume III* (New York: Pinter) 147–175.

Luhmann, N. (1995a), 'Why Does Society Describe Itself as Postmodern', in W. Rasch and C. Wolfe (eds) (2000), *Observing Complexity: Systems Theory and Postmodernity* (Minnesota: University of Minnesota Press).

——. (1995b), *Social Systems* (Stanford: Stanford University Press).

Lumpe, L. (1996), 'Costly Giveaways/ The U.S. Continues to Arm the World, Asking Few Questions and Often Charging Nothing', *Bulletin of the American Scientists* **52** (5), September/October 1996.

——, (2002), 'U.S. Foreign Military Training: Global Range, Global Power, and Oversight Issues', *Foreign Policy in Focus*, Special Report, May 2002, available at <http://www.globalpolicy.org/security/peacekpg/training/us.htm>, accessed 03/03/2006.

Luttwak, E. N. (1996), 'A Post-Heroic Military Policy. (Post-Cold War US Military Policy)', *Foreign Affairs* **75**(4): 33–45.

MacDonald, G. (1999), 'Reform of UK Defense Procurement and State/ Industry Relationships during the 1980s and 1990s', *Defense Analysis* **15**(1): 3–26.

Macedonia, M. (2000), 'Why Digital Entertainment Drives the Need for Speed', *Computer* **33**(2): 124–127.

Machat, M. (2004), 'Bombing in Iraq: Death by Powerpoint', *Airpower* **34**(3): 15–22.

Mackinson, L. (2004), 'Outsourcing the Pentagon: Who Benefits from the Politics and Economics of National Security?', *Center for Public Integrity*, 29 September 2004, available at <http://www.publicintegrity.org/pns/report.aspx?aid=385#9>, accessed 25/05/2005.

Mallet, M. (1974), *Mercenaries and their Masters* (London: The Military Book Society).

Mandel, J. (2006), 'Military Seeks Head Count of Contractors in Iraq', *GovExec.com*, 19 May 2006, available at <http://www.govexec.com/story_page.cfm?articleid=34125&sid=6>, accessed 20/06/2006.

Mandel, R. (2002), *Armies Without States: The Privatization of Security* (London: Lynne Rienner Publisher's Inc.).

Mandelbaum, M. (1998), 'Is Major War Obsolete?' *Survival* **40**(4): 20–38.

Mangold, T and J. Goldberg (2000), *Plague Wars: The Terrifying Reality of Biological Warfare* (New York: St. Martins Press).

Manker, J.E. and K.D. Williams (2004), 'Contractors in Contingency Operations: Panacea or Pain?', *Air Force Journal of Logistics* **28**(3): 12–21.

Marcus, J. (2001), 'Security Concerns Halted Behind the Frontline', *BBC World Service*, 3 December 2001.

Markusen, A. (1997), 'How We Lost the Peace Dividend', *American Prospect* **8**(33): 86–96.

——, (1998), 'The Persistence of Defense Specialised Firms', in G.I. Susman and S. O'Keefe (eds), *The Defense Industry in the Post-Cold War Era: Corporate Strategies and Public Policy Perspectives* (London: Pergamon) 121–146.

——, (2001), 'The Case Against Privatizing National Security', *Meeting of the International Political Science Association*: University of Oklahoma.

Markusen, A. R. and S. S. Costigan (eds) (1999), *Arming the Future: A Defense Industry for the 21st Century* (New York: New York Council on Foreign Relations Press/ Distributed by Brookings Institution Press).

Marren, P. (2003), 'Revolution in Business Affairs?' *Journal of Business Strategy* **24**(4): 4–6.

Matthews, R. and J. Parker (1999), 'Prime Contracting in Major Defense Contracts', *Defense Analysis* **15**(1): 27–42.

Mathiason, N. (2003), 'The First Privatised War: Private Contractors Are Carving Defence Procurement', *The Observer*, 2 March 2003.

Mawdsley, J. (2003), 'The European Union and Defense Industrial Policy', *Bonn International Center for Conversion*, Paper (31).

Maxwell, M.D. (2004), 'The Laws of War and Civilians on the Battlefield/ Are We Undermining Civilian Protections?' *Military Review*, September–October 2004: 17–25.

McCarter, M. (2004), 'ITES Success', *Military Information Technology* **8**(8), available at <http://www.military-information-technology.com/print_article.cfm?DocID=643>, accessed 07/07/2006.

McCarthy, E. (2004), 'Tending to Employees in a War Zone', *Washington Post*, June 10, 2004: E01.

McCarthy, J. (2006), 'Expanding Private Military Sector Faces Structural Change and Scrutiny', *Jane's Intelligence Review*, February 2006: 26–33.

McCullough, J. M. and A. J. Pafford (2002), 'Contractors on the Battlefield: Emerging Issues for Contractor Support in Combat & Contingency Operations', Washington, DC: *US DoD Briefing Papers*, W. Group.

McGhie, S. (2002), 'Private Military Companies: Soldiers, Inc.', *Janes Defence Weekly*, 22 May 2002: 22–26.

McKitrick, J., J. Blackwell, F. Littlepage, G. Kraus, R. Blanchfield, D. Hill (1995), 'Chapter 3: The Revolution in Military Affairs', in L.E. Grinter and B.R. Schneider (eds), *Battlefield of the Future: 21st Century Warfare Issues* (Washington, DC: Air University Press, available at <http://www.airpower.maxwell.af.mil/airchronicles/battle/chp4.html>, accessed 07/07/2006.

McLean, A. (1999), 'PFI in the Sky, or Pie in the Sky? Privatising Military Space', *Space Policy* **15**(1999): 193–198.

McNeill, W.H. (1984), *The Pursuit of Power: Technology, Armed Force, and Society Since 1000 A.D.* (Chicago: University of Chicago Press).

Menarchik, D. (1993), *Powerlift – Getting to Desert Storm: Strategic Transportation and Strategy in the New World Order* (Westport, CT: Praeger).

Mendick, R. (2000), '"Unholy Alliance" to Run Nuclear Weapons Plant', *The Independent Sunday*, 20 February 2000.

Menke, S.M. (2004), 'DOD Sings Satellite Bandwidth Blues', *Government Computer News* **23**(14).

Merindol, V. (2005), 'Defense RDT&E and Knowledge Management: A New Enquiry into Public and Public-Private Coordination', *Defense & Security Analysis* **21**(2): 159–177.

Merle, R. (2003), 'US: Defense Firms Consolidate As War Goes High-Tech', *Washington Post*, 27 May 2003.

Metz, S. (2000), *Armed Conflict in the 21st Century: the Information Revolution and Post-modern Warfare* (Carlisle, PA: Strategic Studies Institute: US Army War College).

Metz, S. and R. A. Millen (2003), *Future War: Future Battlespace the Strategic Role of American Landpower* (Carlisle, PA: Strategic Studies Institute, US Army War College).

Middle East Economic Digest (2004), 'Serco Wins PFI College (Oman)', *MEED Middle East Economic Digest*, 27 February 2004.

Miller, T.C. (2004), 'Army Turns to Private Guards', *Los Angeles Times*, 12 August 2004.

——, (2005), 'U.S. Marines Detained 19 Contractors in Iraq', *L.A. Times*, 8 June 2005.

Miller, W. (2002), 'Dismounted Infantry Takes the Virtual High Ground', *Military Training Technology Online Archive* **7**(8), December 2002.

Moffat, J. (2003), *Complexity Theory and Network-Centric Warfare* (Washington, DC: DoD Command and Control Research Programme, September 2003).

Moller, B. (2002), 'The Revolution in Military Affairs: Myth or Reality?', Copenhagen. *Copenhagen Peace Research Institute*, published on the website *The RMA Debate*, available at <http://www.comw.org/rma/fulltext/second.html>, accessed 01/07/2004.

Monahan, J. (2001), 'Chaos Theory – Modern Military Studies Are No Longer a Case of Orchestral Manoeuvres, Discovers Jerome Monahan', *Guardian Unlimited*, 27 November 2001.

Moskos, C. C., J. A. Williams and D.R. Segal (eds)(2000), *The Postmodern Military: Armed Forces After the Cold War* (Oxford: Oxford University Press).

Mount, I., M. Maier, and D.H. Freedman (2003), 'The New Military Industrial Complex To Arm for Digital-Age War, the Pentagon Has Turned to a New Generation of Defense Contractors. The hardware Is Impressive. It's Also Deadly.', *Business 2.0,* 1 March 2003, available <http://money.cnn.com/magazines/business2/business2_archive/2003/03/01/338099/index.htm>, accessed 16/06/2006.

Mowshowitz, A. (2002), *Virtual Organization: Toward a Theory of Societal Transformation Stimulated by Information Technology* (Westport, CT: Quorum Books).

Mowthorpe, M. (2004), *The Militarization and Weaponization of Space* (Lanham, MD: Lexington Books).

Münkler, H. (2005), *The New Wars* (Cambridge: Polity Press).

Murray, W. (1997), 'Thinking About Revolutions in Military Affairs', *Joint Forces Quarterly*, summer 1997: 69–76.

Murray, W. and R. H. Scales (2003), *The Iraq War: A Military History* (Cambridge. MA: Harvard University Press).

NATO/SHAPE (1999), 'Land Operations in the Year 2020', Neuilly-sur-Seine Cedex, France: *NATO Research and Technology Organization.*

Nederveen Pieterse, J. (2002), 'Globalization at War: War on Terrorism', *International Journal of Peace Studies* **7**(2): 75–91.

——, (2002), 'Globalization, Kitsch and Conflict: Technologies of Work, War and Politics', *Review of International Political Economy* **9**(1): 1–36.

Nichiporuk, B. and C. H. Builder (1995), *Information Technologies and the Future of Land Warfare* (Santa Monica: RAND, Arroyo Center).

Norquist, W.E. (2002), 'How the United States Used Competition to Win the Cold War', *Advances in Competitiveness Research* **10**(1): 1–41.

Northrop Grumman (2004), 'United Kingdom Selects Northrop Grumman as the Preferred Bidder For E-3D AWACS Support Contract Valued at Approximately $1.19 Billion', *Press Release*, 5 August 2004

——, (2005), 'Northrop Grumman Conducts Wargame in Support of Anti-Missile System at JNIC', *Press Release*, 18 February 2005.

Nossal, K. R. (2001), 'Global Governance and National Interests: Regulating Transnational Security Corporations in the Post-Cold War Era', *Melbourne Journal of International Law* **2**, October 2001: 459–476.

O'Brien, K. (1998), 'Military-Advisory Groups and African Security: Privatised Peacekeeping', *International Peacekeeping* **5**(3): 78–105.

——, (2000), 'PMCs, Myths and Mercenaries', *Royal United Service Institute*, Occasional Paper.

O'Hanlon, M. E. (2000), *Technological Change and the Future of Warfare* (Washington, DC: Brookings Institution Press).

Oliver, M. (2006), 'Q & A: Replacing Trident', *Guardian Unlimited*, 15 March 2006.

Onley, D.S. (2005), 'Army Wants to Go Big with Small Computers Program', *Government Computer News* **24**(15), 20 June 2005.

Orsini, E.A. and G.T. Bublitz (1999), 'Contractors on the Battlefield: Risks on the Road Ahead?', *Army Logistician* **31**(1).

Owens, W. A. and E. Offley (2000), *Lifting the Fog of War* (New York: Farrar Straus Giroux).

Paarlberg, R.L. (2004), 'Knowledge as Power: Science, Military Dominance, and U.S. Security', *International Security* **29**(1): 122–151.

Pages, E. (1998), 'Reinventing the Pentagon', in G.I. Susman and S. O'Keefe (eds), *The Defense Industry in the Post-Cold War Era: Corporate Strategies and Public Policy Perspectives* (London: Pergamon) 381–388.

Pagnamenta, R. (2006), 'Defence Companies in Dogfight for MoD Contracts Worth £12 bn', *The Times*, 13 November 2006.

Parker, J. (2000), *Total Surveillance: Investigating the Big Brother World of E-Spies, Eavesdroppers and CCTV* (London: Piatkus Publishers).

Peck, M. (2003), 'Successful War Games Combine Both Civilian and Military Traits', *National Defense Magazine*, November 2003.

Pengelley, R. (2003), 'Future Rapid Effect Systems Leads British Forces' Transformation', *Jane's Defence Industry Review*, 27 August 2003.

Perrow, C. (1999), *Normal Accidents: Living with High-Risk Technologies* (Princeton, NJ: Princeton University Press).

Petersen, L. (2003), 'Outsourcing Government: Service Contracting Has Risen Dramatically in the Last Decade', *The Center for Public Integrity*, 30 October 2003, available at <http://www.publicintegrity.org/wow/report.aspx?aid=68>, accessed 05/07/2006.

Pint, E. M. and R. Hart (2001), *Public-Private Partnerships: Proceedings of the U.S.-U.K. Conference on Military Installation Assets, Operations, and Services* (Santa Monica, CA: RAND, Arroyo Center, April 14–16, 2000).

Pope, C. (2003), 'Boeing Tanker Deal Faces Inquiry', *Seattle Post Intelligencer*, 4 September 2003.

Port, O. (2003), 'Super Soldiers: New Materials and Technologies Could Boost the Mobility and Safety of U.S. Troops', *Business Week*, 28 July 2003: 77–78.

Pryor, M. (1999), 'Digitization, Simulations, and the Future of the Army National Guard', in R.L. Bateman (ed.), *Digital War: A View from the Front Lines* (Novato, CA: Presidio Press).

Rasch, W. and C. Wolfe (2000), *Observing Complexity: Systems Theory and Postmodernity* (Minnesota: University of Minnesota Press).

Rayerman, P. (2003), 'Exploiting Commercial SATCOM: A Better Way', *Parameters US Army War College Quarterly* **33**(4): 54–66.

Raytheon (2002), 'Raytheon Awarded $74 Million Option on US Army Live Training Contract', *Press Release*, 12 November 2002.

Rebar, P. J. (2002), 'Contractor Support on the Battlefield', Carlisle Barracks, PA: *U.S. Army War College*: 1–24.

Repsher Emery, G. (2002), 'Help Wanted: Defence Labs Face Recruitment Challenge', *Washington Technology* **16**(23), 3 April 2002.

Reuters (2005), 'U.S. Military Awards Up to $2.25 Bln in Airlift Work', *Reuters*, 31 August 2005.

Rich, B.R. and L. Janos (2003), *Skunk Works* (London: Time Warner Books).

Richelson, J.T. (1987), *American Espionage and the Soviet Target* (New York: William Morrow&Co.).

——, (1999), *America's Space Sentinels: DSP Satellites and National Security* (Lawrence, KS: University Press of Kansas).

Ricks, T.E. (2001), 'Space Is Playing Field for Newest Wargame: Air Force Exercise Shows Shift in Focus', *Washington Post*, 29 January 2001.

Rifkin, J. (1995), *The End of Work* (New York: Putnam).

Riggs, R. (2004), 'Patriot Missile Problems Worse Than Army Acknowledges', *CBS-11 News*, 11 June 2004.

Rip, M.R. and J.F. Fontanella (1991), 'A Window on the Arab-Israeli "Yom Kippur" War of October 1973: Military Photo-Reconnaissance from High Altitude and Space', *Intelligence and National Security*, Vol. 6, January 1991.

Robinson, L. (2002), 'America's Secret Armies: A Swarm of Private Contractors Bedevils the US Military', *Nation & World*, 11 April 2002.

Roelofs, T. (2007), *Contractor Claims Same Rights as Army*, Newhouse News Service.

Rogers, A. (2003), 'Conflict in the Gulf – VARs Support The Front Lines', *Computer Reseller News*, 24 March 2003.

Rogers, C. (2000), '"Military Revolutions" and "Revolutions in Military Affairs": A Historian's Perspective', in T. Gongora, and H. Von Riekhoff (eds), *Toward a Revolution in Military Affairs? Defense and Security at the Dawn of the Twenty-First Century* (Westport, CT: Greenwood Press: 21–35).

Rogers, P. (2001), 'Towards an Ideal Weapon? Military and Political Implications of the Airborne and Space-based Lasers', *Defence Analysis* **17**(1): 73–88.

Rogerson, K.S. (2000), 'Information Interdependence: Keohane and Nye's Complex Interdependence in the Information Age', *Information, Communication & Society* **3**(3): 415–436.

Romero, S. (2003), 'Military Often Now Enlists Commercial Technology', *The New York Times*, 10 March 2003.

Roosevelt, A. (2003), 'Contractors Follow Military Into Harm's Way', *Navy News & Undersea Technology*: 11.

——, (2004), 'Army Receives Lockheed Martin Aerostat Surveillance System For Iraq', *Defense Daily* **222**(49).

——. (2005), 'Global Defense Cooperation More Vital Than Ever, Krieg Says', *Defense Daily International* **6**(35): 1.

Rowe, D.M. (2005), 'The Tragedy of Liberalism: How Globalisation Caused the First World War', *Security Studies* **14**(3): 407–447.

Ryan, A. (2004), 'Land Forces in the 21st Century Coalition Operations: Implications for the Asia-Pacific', in Evans, R. Parkin and A. Ryan (2004), *Future Armies, Future Challenges: Land Warfare in the Information Age* (Crows Nest, NSW: Allen/Unwin).

SAIC (2005), 'SAIC Awarded Egyptian Armed Forces Training Center Contract', *Press Release*, 7 October 2005.

Salkever, A. (2003), 'The Network Is the Battlefield', *Business Week Online*, 7 January 2003.

Savelyev, A.G. (2004), 'Prospects for US-Russian Cooperation in Ballistic Missile Defense and Outer Space Activities', *Journal of Slavic Military Studies* **17:** 99–109.

Scales, R.H. (ed.) (2000), *Future Warfare Anthology: Revised Edition* (Carlisle Barracks, PA: US Army War College, Collection of Essays).

Schindlemayr, T. (2002), 'Future Personnel: Where Will They Come From?', *Defense & Security Analysis* **18**(1): 85–91.

Schneider, B. R. and L. E. Grinter (1995), *Battlefield of the Future: 21st Century Warfare Issues* (Washington, DC: Air University Press).

Schreier, F. and M. Caparini (2005), *Privatising Security: Law, Practice and Governance of Private Military and Security Companies*, Geneva Centre for Democratic Control of Armed Forces, Occasional Paper No. 6.

Scholl, T.J. (2002), '150 Losing Jobs in Irwin Contractor Switch', *Desert Dispatch*. February 15, 2002.

Schuppe, T.F. (1991), 'Modelling and Simulation: A Department of Defense Critical Technology', *Proceedings of the 1991 Winter Simulation Conference*.

Schwartz, N. D. (2003), 'The Pentagon's Private Army: They Run the Mess Halls. They Program the Weapons. They Even Recruit Soldiers. And If America Goes to War Against Iraq, Private Military Companies Will Play a Bigger Role Than Ever Before', *Fortune* **147**(5): 36–44.

Scott, W.B. (2003), 'Phoenix Air Force: Arizona Co. With Fleet of A-4 Skyhawks Taps Huge Training Market', *Aviation Week & Space Technology* **159**(5): 44–47.

Seebode Waldrop, E. (2004), 'Integration of Military and Civilian Space Assets: Legal and National Security Implications', *Air Force Law Review* **55**(2), spring 2004: 157–232.

Serco (1998), 'Interim Results for the Six Months Ended June 30th 1998', *Serco News*, 2 September 1998.

——, (2003), 'Serco Wins Goose Bay Contract Re-bid', *Press Release*, February, 25, 2003.

Serewicz, L.W. (2002), 'Globalization, Sovereignty and the Military Revolution: From Mercenaries to Private International Security Companies', *International Politics* **39**(1): 75–89.

Serfati, C. (2001), 'The Adaptability of the French Armaments Industry in an Era of Globalization', *Industry and Innovation* **8**(2), August 2001.

Sergounin, A.A. and S.V. Subbotin (1999), *Russian Arms Transfers to East Asia in the 1990s: SIPRI Research Report 15* (Oxford: Oxford University Press).

Shambaugh, D. (2002), *Modernizing China's Military: Problems, Progress, Prospects* (Berkeley: University of California Press).

Shaw, M. (2005), *The New Western Way of War: Risk-Transfer War and its Crisis in Iraq* (Cambridge: Polity Press).

Shearer, D. (1998), *Private Armies and Military Intervention* (Oxford: Oxford University Press) Adelphi Paper (316).

Shepard, S. (1999), 'Soldiers for Hire (Private Military Corporations)', *Contemporary Review* **275**(1603): 66–70.

Sheridan, M. (2003), 'Mess Appeal: Contractors Eagerly Enlist in the Growing Military Market', *Restaurants & Institutions* **113**(3): 57–61.

Shorrock, T. (2005), 'U.S. Deploys Missile Defense System/ Installation Goes Forward Despite Latest Test Failure', *IEEE Spectrum*, February 2005: 10–14.

Shrader, C. R. (1999), 'Contractors on the Battlefield', *Landpower Essay Series*, No. 99-6, May 1999, Association of the United States Army.

Silverstein, K. (2000), *Private Warriors* (New York: Verso).

Singer, J. (2001), 'Air Force War Game Highlights Space Deterrent', *Space.Com*, 29 January 2001.

Singer, P.W. (2002), 'Corporate Warriors: The Rise of the Privatised Military Industry and Its Ramifications for International Security', *International Security* **26**(3): 186–220.

——, (2003), *Corporate Warriors: The Rise of the Privatised Military Industry* (New York: Cornell University Press).

——, (2003b), 'The Enron Pentagon', *The Boston Globe*, 19 October 2003.

——, (2004a), 'Warriors for Hire in Iraq', *Salon.com*, 15 April 2004.

——, (2004b), 'War, Profits, and the Vacuum of Law: Privatised Military Firms and International Law', *Columbia Journal of Transnational Law* **42**(2): 521–550.

——, (2004c), 'Outsourcing the War', *Brookings Institution Foreign Policy Studies*.

——, (2004d), 'The Private Military Industry and Iraq: What Have We Learned and Where to Next?', *Geneva Centre for the Democratic Control of Armed Forces*, Policy Paper.

——, (2005), 'Outsourcing War', *Foreign Affairs* **84**(2), March–April 2005: 119–133.

SIPRI (2005), *SIPRI Yearbook 2005: Armaments, Disarmament and International Security* (Oxford: Oxford University Press).

Skyttner, L. (2001), *General Systems Theory: Ideas and Applications* (River Edge, NJ: World Scientific Publishing).

Sloan, E. C. (2002), *The Revolution in Military Affairs: Implications for Canada and NATO* (Montreal: Queen's University Press).

Smith, C. (2000), 'Russian Mercenaries Flying for Ethiopia: Advisers, Pilots and Artillerymen Engaged in 'Large-Scale Offensive' with Eritrea', *WorldNetDaily*, 18 July 2000.

——, (2002), 'Russian MiGs in Sudan', *NewsMax.com*, January 4, 2002, available at <http://www.newsmax.com/archives/articles/2002/1/4/155909.shtml>, accessed 20/06/2006.

Smith, D. (2005), 'Video Games: Hollywood's New Art', *The Observer*, June 26, 2005.

Smith, E. B. (2002), 'The New Condottieri and US Policy: The Privatization of Conflict and Its Implications', *Parameters US Army War College Quarterly* **32**(4): 104–119.

Smith, M.R. and L. Marx (eds) (1996), *Does Technology Drive History? The Dilemma of Technological Determinism* (Cambridge, MA: The MIT Press).

Snider, D.M. and G.L, Watkins (2000), 'The Future of Army Professionalism: A Need for Renewal and Redefinition', *Parameters US Army War College Quarterly* **30**(3): 5–20.

Spacewar (2005), 'Northrop Grumman Conducts Wargame in Support of Anti-Missile System at JNIC', *Spacewar*, 18 February 2005.

Spearin, C. (2003), 'American Hegemony Incorporated: The Importance and Implications of Military Contractors in Iraq', *Contemporary Security Policy* **24**(3): 26–47.

Stanford, J. (2004), 'Soldiers of Misfortune: Whatever Happened to the Military-Industrial Complex?', *This Magazine* **37**(5): 8–10.

Stanley, R. (1998), 'Employment Effects of Military Restructuring', in M. Kaldor, G. Schméder and U. Albrecht (eds), *Restructuring the Global Military Sector: The End of Military Fordism* (London: Pinter) 35–66.

Stanley-Mitchell, E. A. (2001), 'Technology's Double-Edged Sword: The Case of US Army Battlefield Digitization.', *Defense Analysis* **17**(3): 267–288.

Steinberg, G. (1995), 'Middle East Space Race Gathers Pace', *International Defense Review*, Vol. 28, October 1995.

——, (1998), 'Dual-Use Aspects of Commercial High-Resolution Imaging Satellites', *Mideast Security and Policy Studies* **37**, February 1998.

Strachan, I. (2006), 'Apache Attack Helicopter Training in the British Army', *MS&T* 2006(1): 17–19.

Swartz, L. (2001), 'Overwhelmed by Technology: How Did User Interface Failures On Board the USS Vincennes Lead to 290 Dead?', available at <http://xenon.stanford.edu/~lswartz>, accessed 08/06/2006.

Taibl, P. (1997), 'Outsourcing & Privatization of Defense Infrastructure', *Business Executives for National Security (BENS)*, March 1997, available at <http://www.bens.org/pubs_0397.html>, accessed 02/03/2006.

Talmadge, E. (2004), 'Navy Brings Civilians on Board to Cut Costs', *Associated Press*, 13 May 2004.

Tangredi, S.J. (2003), 'Effects of Globalization on Military Operations', *The European Legacy* **8**(3): 299–315.

Taubman, P. (2003), *Secret Empire: Eisenhower, the CIA, and the Hidden Story of America's Space Espionage* (New York: Simon & Schuster).

Taylor, T. (1999), 'Transatlanticism Versus Regional Consolidation', in D.G. Haglund and S.N. MacFarlane (ed.), *Security, Strategy, and the Economics of Defence Production* (London: McGill-Queens University Press) 66.

——, (2004), 'Contractors on Deployed Operations and Equipment Support', *Defence Studies* **4**(2) 184–198.

Tepperman, J. D. (2002), 'Out of Service - Can Mercenaries Protect Hamid Karzai? (Afghanistan's President)', *The New Republic*, 25 November 2002: 10–14.

Tewes, A. (2006), 'The F-35 Strike Fighter Project: Progress and Issues for Australia', Parliament of Australia, *Parliamentary Library Research Note* **32**, 9 June 2006, available at <http://www.aph.gov.au/library/Pubs/RN/2005-06/06rn32.pdf>, accessed 28/06/2006.

Thomson, J.E. (1996), *Mercenaries, Pirates and Sovereigns: The State-Building and Extraterritorial Violence in Early Modern Europe* (Princeton, NJ: Princeton University Press).

Thornton, J. (2001), '"Maintainability" Drives Fort Worth's Joint Striker Design: Blending Simulation and Ingenuity, Lockheed Martin Up-Ends Design Methods in $320 Billion Program', *Assembly Automation* **21**(3): 204–209.

Tiboni, F. (2004), 'DOD: Contractors Here to Stay: Military, IT Vendors Expect Iraqi-like Job Hazards in Future Wars', *Army Communicator*, summer 2004.

Tiron, R. (2005), 'Merger gives British Firm Bigger U.S. Defense Role', *The Hill*, 29 June 2005.

Tirpak, J. A. (2001), 'Send in the UCAVs: The Air Force Is Pursuing Uninhabited Combat Air Vehicles in a Big Way', *AirForce Magazine* **84**(8), available at <http://www.afa.org/magazine/aug2001/0801ucav.asp>, accessed 07/07/2006.

Todd, P. and J. Bloch (2003), *Global Intelligence: The World's Secret Services Today* (London/New York: Zed Books).

Toffler, A. (1980), *The Third Wave* (New York: Morrow Edition).

Toffler, A. and H. Toffler (1993), *War and Anti-War: Making Sense of Today's Global Chaos* (New York: Warner Books).

Toigo, J.W. (2003), 'Booz Allen Hamilton Inc.', *Washington Technology*, 12 May 2003.

Toomey, C. J. (2003), 'Army Digitization: Making it Ready for Prime Time', *Parameters US Army War College Quarterly* **33**(4): 40–53.

Townsend, M. (2006), 'Army Faces Massive Manpower Shortage/ Government 'Cover-Up' Claimed As Report Reveals Size of Recruitment Crisis', *The Observer*, March 5, 2006.

Traynor, I. (2003), 'The Privatisation of War', *The Guardian*, 10 December 2003.

Trivett, N.R. (1996), 'Air Combat Command Concept of Operations for Theater Battle Management Core Systems', Langley AFB: *US Air Force*, September 30, 1996.

Turkle, S. (2003), 'Technology and Human Vulnerability: A Conversation with MIT's Sherry Turkle', *Harvard Business Review* **81**(9): 43–50.

UK BNSC (2005), 'Successful Launch for Topsat Microsatellite', *British National Space Centre*, Press Release, October 27, 2005, available at <http://www.bnsc.gov.uk/default.aspx?nid=5457>, accessed 15/03/2006.

UK Home Office (2004), 'Strengthening Security, Protecting Identity: Home Office Publishes Identity Cards Bill', London: *UK Home Office Press Release*, 29 November 2004.

UK House of Commons (2002), 'Ninth Report of the Foreign Affairs Committee: Private Military Companies', *House of Commons/ Foreign Affairs Committee*, 23 July 2002, available at <http://www.publications.parliament.uk/pa/cm200102/cmselect/cmfaff/922/92202.htm>, accessed 02/03/2006.

——, (2002), 'Private Military Companies: Options for Regulation', *Foreign and Commonwealth Office*, 12 February 2002, available at <http://www.fco.gov.uk/Files/kfile/mercenaries,0.pdf>, accessed 02/03/2006.

——, (2002), 'UK Defence Committee Press Notice No. 31', 24 October 2002, available at <http://www.parliament.uk/parliamentary_committees/defence_committee/def241002a.cfm>, accessed 29/11/2004.

UK MoD (1998), 'The Strategic Defence Review: Modern Forces for the Modern World', London: *Ministry of Defence*, excerpts available at <http://www.fas.org/nuke/guide/uk/doctrine/sdr98/keypoint.htm>, accessed 20/11/2004.

——, (1999), 'Defence White Paper 1999', London: *Ministry of Defence*, available at <http://www.mod.uk/publications/whitepaper1999/index.htm>, accessed 30/11/2004

——, (2000), 'Kosovo – Lessons from the Crisis', London: *Ministry of Defence*, available at <http://www.kosovo.mod.uk/lessons/contents.htm>, accessed 25/06/2004.

——, (2001), 'Army to Get World's Largest and Most Powerful Tank Transporter', London: *Ministry of Defence*, Press Release, 15 December 2001.

——, (2002a), 'The Strategic Defence Review: A New Chapter: Public Discussion Paper', London: *Ministry of Defence*, available at <http://www.mod.uk/issues/sdr/newchapter.htm>, accessed 10/05/2004.

——, (2002b), 'The Construction of Nuclear Submarine Facilities at Devonport', *Report by the Comptroller and Auditor General*, 6 December 2002

——, (2003a), 'RAF's Latest "Top Gun" Simulators Unveiled', *Ministry of Defence Website*, 26 September 2003, available at <http://www.mod.uk/dpa/news/pn2003/sept03/simulators.htm>, accessed 10/05/2004.

——, (2003b), 'Operations in Iraq: Lessons for the Future', London: *Report of the Ministry of Defence*, 7 July 2003, available at <http://www.mod.uk/publications/iraq_futurelessons/index.html>, accessed 12/05/2004.

——, (2003), Ministry of Defence Consolidated Departmental Resource Accounts 2001-02/ Private Finance Initiative (PFI) Commitments, available at <www.archive2.official-documents.co.uk/document/deps/hc/hc47/4737.htm>, accessed 10/05/2004.

——, (2004a), 'Policy Paper: Paper Number 4 Defence Acquisition', London: *Ministry of Defence*, available at <http://www.mod.uk/linked_files/def_acquisition.pdf>, accessed 20/11/2004.

——, (2004b), 'The Government Expenditure Plans 2004/ 2005', London: *Ministry of Defence*, available at <http://www.mod.uk/linked_files/publications/gep_04-05.pdf>, accessed 29/11/2004.

——, (2004c), 'MoD Awards New Partnering Deal For Logistics Support', *Press Release*, 3 February 2004.

——, (2005), 'Defence Logistics Organisation: The 2005 Plan', London: *Ministry of Defence*, available at <http://www.mod.uk/linked_files/dlo/policy_and_strategy/dloplan2005.pdf>, accessed 25/08/2005.

UK MoD/ABRO (2005), 'ABRO and BAE SYSTEMS Land Systems Form Strategic Partnership', London: *ABRO*, Press Release, 8 February 2005.

UK MoD (2006a), 'New Chinook Support Arrangements Will Save £170 Million', London: *Defence News*, Press Release, 16 February 2006, available at <http://www.mod.uk/DefenceInternet/DefenceNews/EquipmentAndLogistics/NewChinookSupportArrangementsWillSave170m.htm>, accessed 16/03/2006.

——, (2006b), 'TSP 6: Global Location of UK Regular Forces', *UK MoD Defence Analytical Services Agency*, March 16, 2006, available at <http://www.dasa.mod.uk>, accessed 16/03/2006.

UK National Audit Office (2002a), 'Ministry of Defence: Major Projects Report 2002', *National Audit Office*, 4 December 2002, available at <http://www.nao.org.uk/publications/ nao_reports/02-03/020391.pdf>, accessed 02/03/2006.

——, (2002b), 'The Construction of Nuclear Submarine Facilities at Devonport', *Report by the Comptroller and Auditor General*, 6 December 2002.

——, (2002c), 'Major Repair and Overhaul of Land Equipment', *Report by the Comptroller and Auditor General*, 26 April 2002.

——, (2004), 'Ministry of Defence Battlefield Helicopters', *Report by the Comptroller and Auditor General*, April 7, 2004, available at <http://www.nao.org.uk/publications/ nao_reports/03-04/0304486.pdf>, accessed 02/03/2006.

UK National Audit Office/MoD, 'Training New Pilots: A Report by the Comptroller and Auditor General', available at <http://www.ukmfts.mod.uk/files/pilots.pdf>, accessed on 29/11/2004.

Ullman, H. (1996), *Shock and Awe: Achieving Rapid Dominance* (Washington, DC: Center for Advanced Concepts).

Urry, J. (2003), *Global Complexity* (Cambridge: Polity Press).

US Air Force (1990), 'The Air Force and U.S. National Security: Global Reach-Global Power', Washington, DC: *US Air Force Department.*

US Congressional Budget Office/D. Clay-Mendez (1995), 'Public and Private Roles in Maintaining Military Equipment at the Depot Level', *Congressional Budget Office*, July 1995.

——, (1997), 'Moving U.S. Forces: Options for Strategic Mobility', *Congressional Budget Office*, February 1997, available at <http://www.cbo.gov/showdoc.cfm?index=11&sequence=0>, accessed 02/03/2006.

——, (2000), 'Budget Options for National Defense', Washington, DC: *Congressional Budget Office*, available at <http://www.cbo.gov/showdoc.cfm?index=1873&sequence=0>, accessed 02/03/2006.

US DoD/S. Glasstone and P.J. Dolan (1977), *The Effects of Nuclear Weapons*, Washington, DC: US DoD Energy Research and Development Administration, available at <http://www.princeton.edu/~globsec/ publications/effects/effects3.pdf>, accessed 09/07/2006.

US DoD (1996), 'Improving the Combat Edge Through Outsourcing', *Defense Issues* **11**(30): 16.

——, (1997), 'The Report of the Quadrennial Defense Review', Washington, DC, May 1997, available at <http://www.defenselink.mil/pubs/qdr/>, accessed 08/08/2004.

US DoD/J. Gansler (1998), 'Modeling and Simulation: Designing Affordable Weapons Systems For the 21s Century', Washington, DC: *DMSO Modeling and Simulation Briefing*, Keynote Speech, 2 June 1998.

US DoD. (1999), 'Final Report of the Defense Science Board Task Force on Globalization and Security', Washington, DC: *Office of the Undersecretary of Defense for Acquisition and Technology*, December 1999.

——, (2000), 'Report of the Defense Science Board Task Force on Defense Software', Washington, DC: *Office of the Under Secretary of Defense for Acquisition and Technology*, November 2000.

——, (2001a), 'Quadrennial Defense Review Report', Washington, DC: *Office of the Secretary of Defense*, 30 September 2001.

——, (2001b), 'Report of the Defense Science Board Task Force on Training Superiority & Training Surprise', Washington, DC: *Office of the Under Secretary of Defense for Acquisition and Technology*, January 2001.

——, (2002), 'Nuclear Posture Review: Excerpts', available at <http://www.globalsecurity.org/wmd/library/policy/dod/npr.htm>, accessed 20/06/2006.

——, (2003a), *Directive 5000.1*, May 12, 2003.

——, (2003b), 'Annual Base Structure Report/ A Summary Of DoD Real Property Inventory', Washington, DC: *Office of Deputy Undersecretary of Defense (Installations and Environment)*, available at <http://www.defenselink.mil/news/Jun2003/basestructure2003.pdf>, accessed 15/03/2006.

——, (2004a), 'DoD Maintenance Policy, Programs and Resources/ Fact Book 2004', Washington, DC: *Office of the Deputy Under Secretary of Defense* (Logistics and Material Readiness).

——, (2004b), 'Worldwide Manpower Distribution by Geographic Area', Washington, DC: *Department of Defense Washington Headquarters Services Information Technology and Management Division*, 30 September 2004, available at <http://www.dior.whs.mil/mmid/pubs.htm#M05>, accessed 15/03/2006.

US General Accounting Office (1992), 'The Performance of the Patriot Missile in the Gulf War', *GAO Report prepared fot the Government Operations Committee, House of Representatives*, October 1992.

——, (1997), 'Contingency Operations: Opportunities to Improve the Logistics Civil Augmentation Program', *Report to Congressional Requesters*, February 1997, available at <http://www.gao.gov/cgi-bin/getrpt?GAO/NSIAD-97-63>, accessed 02/03/2006.

——, (1998), 'Public-Private Competitions: Access On Records Is Inhibiting Work On Congressional Mandates', *Testimony Before the Subcommittee on Readiness, Committee on Armed Services, U.S. Senate*, 4 March 1998, available at <http://www.access.gpo.gov/cgi-bin/getdoc. cgi?dbname=gao&docid=f:ns98111t.txt.pdf>, accessed 02/03/2006.

——, (2000), 'Gulf War Illnesses/ Management Actions Needed to Answer Basic Research Questions', *Report to the Chairman, Subcommittee on National Security, Veterans' Affairs and International Relations, Committee on Governmental Reform, House of Representatives*, January 2000, available at <http://www.gao.gov/new.items/d04767.pdf>, accessed 02/03/2006.

——, (2001), 'Defense Maintenance: Sustaining Readiness Support Capabilities Requires a Comprehensive Plan', *General Accounting Office*, March 23, 2001.

——, (2003), 'Military Operations: Contractors Provide Vital Services to Deployed Forces But Are Not Adequately Addressed in DoD Plans', *Report to the Subcommittee on Readiness and Management Support, Committee on Armed Services, U.S. Senate*, June 2003.

——, (2004), 'Uncertainties Remain Concerning the Airborne Laser's Costs and Military Utility', 19 May 2004, available at <http://www.gao.gov/htext/d04643r.html>, accessed 22/10/2005.

——, (2005), 'Defense Logistics: DoD Has Begun to Improve Supply Distribution Operations, but Further Actions Are Needed to Sustain These Efforts', *Report to the Subcommittee on Readiness, Committee on Armed Services, House of Representatives*, August 2005.

——, (2006), 'Rebuilding Iraq: Actions Still Needed to Improve the Use of Private Security Providers', Testimony Before the Subcommittee on National Security, Emerging Threats, and International Relations, Committee on Government Reform, 13 June 2006.

US Joint Chiefs of Staff (1995), 'Joint vision 2010', Washington, DC: *U.S. Department of Defense*.

——, (2000), 'Joint Vision 2020', Washington, DC: *U.S. Department of Defense*.

——, (2004), 'National Military Strategy of the United States', available at <http://www.defenselink.mil/news/Mar2005/d20050318nms.pdf>, accessed 17.08.05.

US Marine Corps (1996), 'MCDP-6: Command and Control', Washington, DC: *Department of the Navy: Headquarters of the Marine Corps*.

US Navy (1994), 'Forward ... From the Sea', Washington, DC: *US Navy Department*.

US Space Command (1997), 'Vision for 2020', Peterson AFB: *US Space Command*.

Uttley, M.R.H. (2004), 'Private Contractors on Deployed Operations: The United Kingdom Experience', *Defence Studies* **4**(2): 145–165.

Van Creveld, M.L. (1979), *Supplying War: Logistics from Wallenstein to Patton* (Cambridge: Cambridge University Press).

——, (1985), *Command in War* (Cambridge, MA: Harvard University Press).

——, (1989), *Technology and War: From 2000 B.C. to the Present* (London: Collier Macmillan).

——, (1991), *The Transformation of War* (Toronto/New York: Collier Macmillan).

——, (1996), 'The Fate of the State', *Parameters US Army War College Quarterly* **26**(1): 4–18.

——, (1999), *The Rise and Decline of the State* (Cambridge: Cambridge University Press).

Verton, D. (2005), 'Simulating Fallujah/ Graphics Engines, Supercomputers and Real Gunpowder', *Computerworld*, 31 January 2005: 28.

Von Hoffman, N. (2004), 'Contract Killers: How Privatizing the US Military Subverts Public Oversight', *Harpers Magazine* **308**(1849): 79–81.

Wait, P. (2003), 'Contractors Support Systems on Front Lines', *Government Computer News* **22**(6): 8.

Wakeman, N. (2004), 'Industry Titans Still Blazing Trail', *Washington Technology* **19**(11), 30 August 2004.

Waldrop, M.M. (1992), *Complexity: The Emerging Science at the Edge of Chaos* (London: Penguin).

Wall, R. and D.A. Fulghum (2004), 'Failure to Perform', *Aviation Week & Space Technology* **161**(24), 20 December 2004.

Warden, J.A. III (1995), 'Air Theory for the Twenty-first Century', in L.E. Grinter and B.R. Schneider (eds), *Battlefield of the Future: 21st Century Warfare Issues*

(Washington, DC: Air University Press, available at <http://www.airpower.maxwell.af.mil/airchronicles/battle/chp4.html>, accessed 07/07/2006.

Wayne, L. (2002), 'America's For Profit Secret Army', *New York Times*, 13 October 2002.

Weiner, T. (2004), 'Lockheed and the Future of Warfare', *New York Times*, 28 November 2004.

Weinberger, S. (2006), *Imaginary Weapons: A Journey through the Pentagon's Scientific Underground* (New York: Nation Books).

West, N. (1998), *The Secret War for the Falklands: SAS, MI6 and the War Whitehall Nearly Lost* (New York: Time Warner Books).

Whyte, D. (2003), 'Lethal Regulation: State-Corporate Crime and the United Kingdom Government's New Mercenaries', *Journal of Law and Society* **30**(4): 575–600.

Williams, R. (2004), 'DoD Takes Personnel Recovery to Next Level', *DefenseLINK News*.

Wilson, I. (2005), '"Providing for the Common Defense?" The Outsourcing of America's National Security through the Commercial Sale of Arms and Arms Production', *Democracy and Security* **1**: 73–104.

Wilson, J.R. (2004), 'Simulation Is No Longer Just for the Laboratory', *Military & Aerospace Electronics*, January 2004: 16–20.

Wilson, J. (2005), 'US Lowers Standards in US Army Number Crisis', *The Guardian*, 4 June 2005.

Wirbel, L. (2002), 'The Space Industry: Supporting U.S. Supremacy', *Foreign Policy in Focus* **7**(13), November 2002.

——, (2002b), 'Star Wars: Imperialism in Space', in C. Boggs (ed.), *Masters of War/Militarism and Blowback in the Era of American Empire* (London: Routledge) 59–82.

——, (2004), *Star Wars: US Tools of Space Supremacy* (London: Pluto Press).

Witte, G. (2006), 'Army to End Expensive, Exclusive Halliburton Deal', *Washington Post*. 12 July 2006: A01.

Wolf, J. (1999), 'Lockheed Martin Wins NORAD Upgrade Contract', *Reuters*, 19 September 1999.

Wood, D. (2003), 'Some of Army's Civilian Contractors Are No-Shows in Iraq', *Newhouse News Service*, 31 July 2003.

Wriston, W.B. (1997), 'Bits, Bytes and Diplomacy', *Foreign Affairs* **76**(5): 172–182.

Yeoman, B. (2003), 'Soldiers of Good Fortune: They Fly Helicopters, Guard Military Bases, and Provide Reconnaissance. They're Private Military Companies - and They're Replacing U.S. Soldiers in the War on Terrorism', *Mother Jones* **28**(3): 38–47.

Zaloga, S. (2000), 'Russia's Strategic Forces Stumble', *Jane's Intelligence Review* **12**(10): 12–14.

Zamparelli, S. (1999), 'Contractors on the Battlefield: What Have We Signed Up For?', *Air Force Journal of Logistics* **23**(3): 8–17.

Zucker, J.S. (2004), 'The Boeing Suspension: Has Increased Consolidation Tied the DoD's Hands?', *Army Lawyer*, April 2004.

Index